THE US ARMED FORCES OFFICER

THE US ARMED FORCES OFFICER

ESSAYS ON LEADERSHIP, COMMAND, OATH, AND SERVICE IDENTITY

RICHARD M. SWAIN AND
ALBERT C. PIERCE

Foreword by General Joseph F. Dunford, Jr.
Preface by Major General Frederick M. Padilla

Skyhorse Publishing

Originally published in the United States by National Defense University Press

10 9 8 7 6 5 4 3 2 1

Library of Congress Cataloging-in-Publication Data is available on file.

Cover design by Paul Qualcom

Print ISBN: 978-1-5107-4308-3
Ebook ISBN: 978-1-5107-4315-1

Printed in the United States of America

Contents

Foreword

In 1950, the great Soldier-Statesman George C. Marshall, then serving as the Secretary of Defense, signed a cover page for a new book titled *The Armed Forces Officer*. That original version of this book was written by none other than S.L.A. Marshall, who later explained that Secretary Marshall had "inspired the undertaking due to his personal conviction that American military officers, of whatever service, should share common ground ethically and morally." Written at the dawn of the nuclear age and the emergence of the Cold War, it addressed an officer corps tasked with developing a strategy of nuclear deterrence, facing unprecedented deployments, and adapting to the creation of the Department of Defense and other new organizations necessary to manage the threats of a new global order.

Now, in the second decade of the 21st century, our nation is again confronted with a volatile and complex security environment, and addressing the challenges of our time will place new demands on military leaders at all levels. We in the Profession of Arms will continue to adapt our training and education programs, as we have always done, to provide our officers with the intellectual and practical tools necessary to succeed in this unpredictable and unstable world.

The character of warfare may change over time, but its nature does not. As novel as much of the current security environment may seem, George C. Marshall's wisdom still rings true today. Regardless of the challenges we face, our leaders, especially our officers, must share a moral foundation and practice a common professional ethic. Our tactics, techniques, and practices may change, but our bedrock principles remain the same.

This new edition of *The Armed Forces Officer* articulates the ethical and moral underpinnings at the core of our profession. The special trust and confidence placed in us by the Nation we protect is built upon this foundation. I commend members of our officer corps to embrace the principles of this important book and practice them daily in the performance of your duties. More importantly, I expect you to imbue these values in the next generation of leaders.

—Joseph F. Dunford, Jr.

General, U.S. Marine Corps
Chairman of the Joint Chiefs of Staff

Preface

In 2007, the National Defense University and the NDU Press published a new edition of *The Armed Forces Officer.* That book was written in the period from 2002 to 2005 as a 21st-century version of a work originally published by the Office of the Secretary of Defense in 1950. Three subsequent editions followed throughout the last half of the 20th century. The 2007 edition was drafted by representatives of the national Service academies, with additional contributions by the Marine Corps University.

A few years ago, senior leaders in the Department of Defense decided that the times called for a new edition of the book. To accomplish this task, NDU turned to two people who had played key roles in writing and editing the 2007 version, Dr. Albert C. Pierce, NDU's Professor of Ethics and National Security, and Dr. Richard Swain, a retired Army colonel and former Professor of Officership at the U.S. Military Academy at West Point. Together, they produced this 2016 edition of *The Armed Forces Officer.*

NDU is proud to publish this new book as part of its ongoing efforts in "Educating, Developing, and Inspiring National Security Leadership."

—Frederick M. Padilla

Major General, U.S. Marine Corps
President, National Defense University

THE ARMED FORCES OFFICER

CHAPTER ONE

The Commission and the Oath

You become an officer in the Armed Forces of the United States by accepting a commission and swearing the oath of support for the Constitution required by Article VI of "all executive and judicial Officers [the President excepted], both of the United States and of the several states."[1] The commission and the oath constitute an individual moral commitment and common ethical instruction. They legitimize the officer's trade and provide the basis of the shared ethic of commissioned leadership that binds the American military into an effective and loyal fighting force. They are the foundation of the trust safely placed in the Armed Forces by the American people. The commission and oath unite all Armed Forces officers in a common undertaking of service to the Nation.

The Commission

Though the paths taken to the tender of a Federal military commission are various, the form of the document is common among the Armed Forces, save for the fact that each reflects appointment in a particular branch of the Armed Forces (Army, Marine Corps, Navy, Air Force, or Coast Guard). The commission is granted under the President's powers in Article II, Section 2, of the Constitution. It is a notice of appointment, a grant of executive authority, and an admonition for obedience. It is bestowed, the commission says, because of the "special trust and confidence" reposed by the President "in the patriotism, valor, fidelity and abilities" of the appointee. The officer is enjoined to "carefully and diligently discharge the duties" of his or her office. Subordinates are charged to render the obedience due an officer of his or her station. The officer is admonished to "observe and follow such orders and directions . . . as may be given by" the President or the President's successors, "or other Superior Officers *acting in accordance with the laws of the United States of America* [emphasis added]."[2] No grant of professional discretion exempts any Armed Forces officer from the obligation to act within the confines of the law.

The form of the commission document remains much like that granted by the Continental Congress to officers of the Continental Army during the American Revolution.[3] The wording of the current commission replaces the 1777 *conduct* with *abilities*.[4] By way of comparison, Article I of the 1775 "Rules for the Regulation of the Navy of the United States Colonies of North America" reads: "The Commanders of all ships and vessels belonging to the thirteen United Colonies are strictly required to show themselves a good example of honor and virtue to their officers and men."[5]

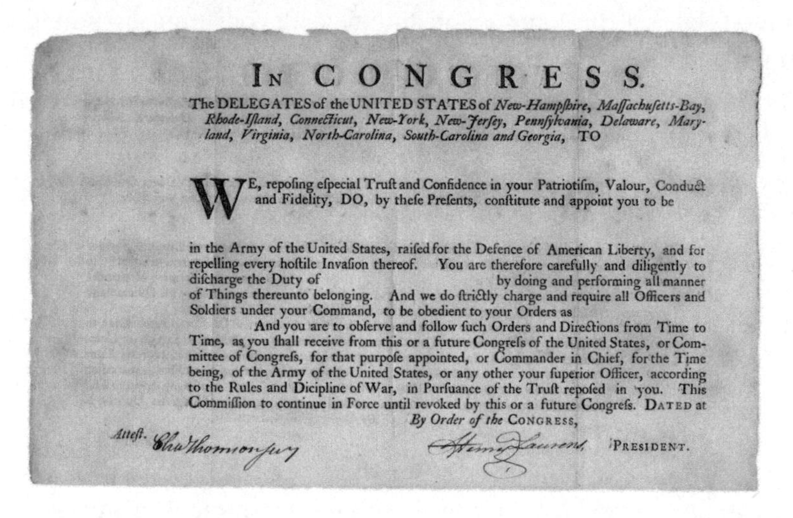

IN CONGRESS.

The DELEGATES of the UNITED STATES of *New-Hampſhire, Maſſachuſetts-Bay, Rhode-Iſland, Connecticut, New-York, New-Jerſey, Pennſylvania, Delaware, Maryland, Virginia, North-Carolina, South-Carolina and Georgia*, TO

WE, repoſing eſpecial Truſt and Confidence in your Patriotiſm, Valour, Conduct and Fidelity, DO, by theſe Preſents, conſtitute and appoint you to be

in the Army of the United States, raiſed for the Defence of American Liberty, and for repelling every hoſtile Invaſion thereof. You are therefore carefully and diligently to diſcharge the Duty of by doing and performing all manner of Things thereunto belonging. And we do ſtrictly charge and require all Officers and Soldiers under your Command, to be obedient to your Orders as

And you are to obſerve and follow ſuch Orders and Directions from Time to Time, as you ſhall receive from this or a future Congreſs of the United States, or Committee of Congreſs, for that purpoſe appointed, or Commander in Chief, for the Time being, of the Army of the United States, or any other your ſuperior Officer, according to the Rules and Dicipline of War, in Purſuance of the Truſt repoſed in you. This Commiſſion to continue in Force until revoked by this or a future Congreſs. DATED at

By Order of the CONGRESS,

Atteſt. PRESIDENT.

The Armed Forces of the United States depend for their success, indeed for their existence, on a web of trust beginning with that between them and the American people and their government. The President expects the officer to live up to the expectations expressed in the commission. The people depend upon the Armed Forces for their security in a dangerous world. They provide their sons and daughters as Soldiers, Marines, Sailors, Airmen, and Coastguardsmen, in trust that their lives will be risked reluctantly and expended parsimoniously only as required for important tasks. They expect the leaders and members of their Armed Forces to be both effective and accountable before the law and public opinion. The people pay their taxes in order to ensure the safety of the Nation. Notably, providing for "the common defense" precedes promoting "the general welfare" as a founding purpose in the preamble of the Constitution.

In return for their investment, the American people expect reliable, effective, honorable, and efficient performance. They demand military leaders who demonstrate high standards of character and competence and who conduct themselves in a manner that reflects basic principles of integrity, justice, and fairness toward all subordinates.[6] When these expectations are disappointed, the people and their government withhold the trust, resources, and discretionary latitude the Services enjoy in more normal times. Equally important, when a lack of public trust becomes evident, the morale of Servicemembers suffers. Military men and women question the value of their sacrifices, the worth of their cause, and trust in their leaders. Discipline becomes problematic.[7]

In 1950, S.L.A. Marshall began the first edition of *The Armed Forces Officer* with a chapter titled "The Meaning of Your Commission." His opening sentences read as follows:

> *Upon being commissioned in the Armed Services of the United States, a man incurs a lasting obligation to cherish and protect his country and to develop within himself that capacity and reserve strength which will enable him to serve its arms and the welfare of his fellow Americans with increasing wisdom, diligence, and patriotic conviction. This is the meaning of his commission.*[8]

Lingering over the implications of the four virtues to which the President attests, Marshall gave pride of place to *fidelity*, discounting patriotism, which he had largely defined in his opening sentence because, he said, it could be assumed. *Valor* he set aside because it remained unknown until it was tested. *Abilities* depended on individual nature. "Fidelity," he asserted, "is the derivative of personal decision . . . the jewel within reach of every man who has the will to possess it."[9]

Patriotism, the zealous devotion to one's own country, is a suspect virtue today, more credible when recognized by others than when self-proclaimed. Samuel Johnson's assertion that patriotism represents the last refuge of the scoundrel seems too often justified in the conduct of the professionally patriotic. If experience teaches the unwisdom of Marshall's too easy presumption of patriotism by the officer, its recognition by others—the public and those with whom one has influence—should remain an important aspiration of every Armed Forces officer. Evident and motivating love of country is the beginning of authority's legitimacy.

Valor represents the virtue, or quality of mind, that enables a person to face danger with boldness or firmness. It is an essential if not sufficient requirement of any who would aspire to lead those intended to go into harm's way. Marshall may have been correct that valor remains unknown until tested. If so, officers would do well to examine themselves to the extent practicable and, by repeated experience and reflection, gain confidence in their own measure.

On the other hand, contrary to Marshall's view, abilities, or the "power or skill to do something," are subject to training and improvement. Abilities can be enhanced. Abilities become capabilities or capacities through practice and application. Demonstrated abilities, not least a certain athleticism for what is a physically demanding calling, may be the basis for initial commissioning, but the officer remains under obligation to extend his or her inherent abilities to their maximum potential. Perhaps the ability of intellectual growth is the most important, which returns us to Marshall's foremost virtue, fidelity.

Of the four commissioning virtues, Marshall preferred fidelity because he saw that it was a matter of individual choice or will. The *Oxford English Dictionary* defines fidelity as "the quality of being faithful; faithfulness, loyalty, *unswerving allegiance* to a person, party, bond, etc. [emphasis added]." Fidelity is the foundation of the various strands of trust that mark the relationships of the Armed Forces officer. It means the officer will stand fast in the face of hardship and danger. Fidelity, faithfulness to the Constitution, binds the officer to the Nation and the people the officer serves. Fidelity to the Service, and to those in superior command, ensures discipline and reliability. Fidelity to the men and women entrusted to the officer's care is the basis of *esprit* and collective performance. At its most basic level, it is acceptance of the primacy of duty in all things.

The Oath

Acceptance of the commission is conditional upon execution of the constitutional oath. The commissioning oath is an individual commitment, made freely, publicly, and without mental reservation, to support and defend the compact that forms the United States "against all enemies foreign and domestic; to bear true faith and allegiance to the same," and, echoing the commission, to "well and faithfully discharge the duties of the office on which I am about to enter."[10] The current form of the officer's oath is found in Title 5, "Government Organization and Employees," of the U.S. Code; the enlisted member's oath of enlistment is in Title 10, "Armed Forces." The form of the oath has changed over time, most notably during the Civil War and its aftermath, as the Congress of the United States sought to protect itself, in the first place, from a repeat of officers and officials "going South," and also to keep former Confederate officials out of government.[11]

Marshall had less to say about the oath of office than the commission, though he observed in his discussion of *esprit* that

> *the interesting and important thing that happens to a man when he enters military service is that, the moment he takes the oath, loyalty to the arms he bears ranks first on the list, above all other loyalties. . . . In his life, service to country is no longer a beautiful abstraction; it is the sternly concrete and unremitting obligation of service to the regiment, the group or the ship's company. . . . In this radical reorientation of the individual life and the arbitrary imposition of a commanding loyalty is to be found the key to the esprit of any military organization.*[12]

There are stark differences between the undertaking of the civil servant who subscribes to the constitutional oath and the military officer who does the same to activate a military commission. Notably, while both the commission and oath involve, on the one hand, the admonition for careful and diligent discharge of duties and, on the other, a commitment "to well and faithfully discharge the duties of the office," both are silent as to what those duties might encompass beyond the shared purpose of protecting and defending the Constitution. But it is precisely the nature of the task that makes Armed Forces officers unique among executive officers of the government.

The military oath is implicitly a commitment to what a British general, Sir John Hackett, called "the ordered application of force under an unlimited liability."[13] The military man or woman may be called upon at any time to perform duties under conditions not only of great discomfort, but also of threat of serious injury, loss of limb, or death. Officers' particular duty, or at least that which defines their corps, is the leadership and direction of men and women in the disciplined use of lethal force (or the threat thereof), in the pursuit of purposes sanctioned by the state and legal under the Constitution and in international law. The Supreme Court of the United States has observed: "An army is not a deliberative body. It is an executive arm. Its law is that of obedience. No question can be left open as to the right of command in the officer, or the duty of obedience in the soldier."[14]

The "Otherness" of Officers

Not only in the United States, but also within the armed forces of other established nation-states, the officer corps generally exists as a body apart from the enlisted force. Commissioned officers are intentionally different. General Hackett observed that, to underscore the officer's right to command:

> *there is in armies a tendency to set up an officer group with an otherness as a step towards or if necessary even in some degree a replacement of, the betterness you require. The officer is set apart, clothed differently and given distinguishing marks. His greater responsibilities are rewarded with greater privileges. There is some insistence on a show of respect. He is removed from that intimate contact with the men under his command which can throw such a strain upon the relationship of subordination.*[15]

In the United States, Armed Forces officers are set apart as a group within the wider profession of arms: in uniform, insignia, formal respect required, authority assigned, responsibility, and limitations on appropriate interaction with other members. The commission document, the unique form of appointment, is one of these distinguishing features and has already been addressed. The salute is a required greeting of senior officers, rendered by subordinates, enlisted and commissioned alike; likewise, the use of certain forms of address, such as *Sir* or *Ma'am*, is an obligatory sign of respect. An officer's authority is underwritten by the Uniform Code of Military Justice in the severity of punishment for offenses committed against commissioned officers in execution of their office.[16] Indicative of differences in responsibility, there are offenses in the Uniform Code of Military Justice that only an officer can commit; most notably, these are Article 88, "Contempt for officials," and Article 133, "Conduct unbecoming an officer and a gentleman" (one of the few remaining couplings of the terms "officer" and "gentleman").

Among the U.S. Armed Forces, the Marine Corps is the most eloquent in defining the "otherness of officers." In the *Marine Corps Manual*, the core document of the Marine Corps, paragraph 1100, "Leadership," includes the following:

> *The special trust and confidence, which is expressly reposed in officers by their commission, is the distinguishing privilege of the officer corps. It is the policy of the Marine Corps that this privilege be tangible and real; it is the corresponding obligation of the officer corps that it be wholly deserved.*
>
> *(1) As an accompanying condition commanders will impress upon all subordinate officers the fact that the presumption of integrity, good manners, sound judgment, and discretion, which is the basis for the special trust and confidence reposed in each officer, is jeopardized by the slightest transgression on the part of any member of the officer corps. Any offense, however minor, will be dealt with promptly, and with sufficient severity to impress on the officer at fault, and on the officer corps. Dedication to the basic elements of special trust and confidence is a Marine officer's obligation to the officer corps as a whole, and transcends the bonds of personal friendship.*
>
> *(2) As a further and continuing action, commanders are requested to bring to the attention of higher authority, referencing this paragraph, any situation, policy, directive, or procedure*

which contravenes the spirit of this paragraph, and which is not susceptible to local correction.[17]

Whereas General Hackett looked to institutional "otherness" to act in place of a "betterness" that justified command, the American Brigadier General S.L.A. Marshall saw the origin of the officer's prestige as derived from the "exceptional and unremitting responsibility," which is his or her lot, and he saw that the importance of this esteem and trust was one of the reasons the Services placed emphasis on personal honor. "They know," Marshall wrote, "that the future of our arms and the well-being of our people depend upon a constant renewing and strengthening of public faith in the virtue of the [officer] corps."[18] Marshall observed as well that "while he continues to serve honorably, it [the Nation] will sustain him and *will clothe him with its dignity* [emphasis added]."[19] In short, the Nation will bestow on the officer the authority to command his or her fellow citizens.

The Army tends to follow Marshall, adding to responsibility the attribute of specialized knowledge, and writes the following in its leadership reference manual:

> *2-6. Officers are essential to the Army's organization to command units, establish policy, and manage resources while balancing risks and caring for their people and families. . . .*
> *2-7. Serving as an officer differs from other forms of Army leadership by the quality and breadth of expert knowledge required, in the measure of responsibility attached, and in the magnitude of the consequences of inaction or ineffectiveness. . . . While officers depend on the counsel, technical skill, maturity, and experience of subordinates to translate their orders into action, the ultimate responsibility for mission success or failure resides with the officer in charge.*[20]

All the Services have explicit policies on improper relationships, or *fraternization*, between ranks, intended to maintain good order and discipline by forbidding certain transactions and relationships between the different classes of membership.

Society's respect for the professional officer is conditioned on reliable, effective, honorable, and efficient performance of duty. As

General William T. Sherman warned officers attending the new School of Application for Infantry and Cavalry at Fort Leavenworth in October 1882:

> *No other profession holds out to the worthy so certain a reward for intelligence and fidelity, no people on earth so generously and willingly accord to the soldier the most exalted praise for heroic conduct in action, or for long and faithful service, as do the people of the United States; nor does any other people so overwhelmingly cast away those who fail at the critical moment, or who betray their trusts.*[21]

The military ethic is a warrior ethic and the military ethos is a warrior ethos, a point made clear by the Soldier's Creed, the Marine Hymn, the Sailor's Creed, and the Airman's Creed. This seems unlikely to change, even in the era of cyber-conflict and unmanned attack aircraft. In 2003, American journalist William Pfaff wrote in his essay "The Honorable Absurdity of the Soldier's Role" that the soldier's lot "is inherently and voluntarily a tragic role, an undertaking to offer one's life, and to assume the right to take the lives of others. . . . The intelligent soldier recognizes that the two undertakings are connected. His warrant to kill is integrally related to his willingness to die."[22] When one is not *willing* to go into harm's way, he or she is not a soldier but a technician of death, or just a technician. A defining moral quality is absent. The military ethic is based on a commitment to disciplined service under conditions of unlimited liability, whether or not one has a military occupational specialty that involves combat.[23]

The year following Pfaff's essay, in an important and eloquent public strategy document, the Secretary of the Army and the Army Chief of Staff reflected that this right to take the lives of others involves a burden of discrimination—that "Only the true warrior ethos can moderate war's inevitable brutality."[24] Later that year, Lieutenant General James Mattis, USMC, perhaps the most admired warrior-general of the day, told an audience of Naval Academy Midshipmen: "The first thing, my fine young men and women, you must make certain that your troops know where you are coming from and what you stand for and, more importantly, what you will not tolerate."[25] Mattis went on to recount a story brought to him by a non-U.S. war correspondent who witnessed the actions of a young Marine rifleman in Mamadia, Iraq. The correspondent, skeptical of the Marine Corps admonition of "No better friend, no worse enemy," had observed the young Marine protecting nearby Iraqi civilians caught in the danger zone while simultaneously fighting off terrorists. "Now think," Mattis concluded, "what that says about a 19-year-old who could discriminate."[26]

What the American people expect from their Armed Forces, and trust that they will receive, is reliable, effective, honorable, and efficient service, whenever, wherever, and in whatever form the government of the day decides is necessary. The guarantee of that service is internalization in every officer of the expectations embodied in the commission and the oath: patriotism, valor, fidelity, and abilities; dedication to the protection of the letter of and the values embodied in the Constitution; and a willingness to offer, if required, what President Lincoln called "the last full measure of devotion" in its defense. In the first Raymond Spruance Lecture at the Naval War College, Herman Wouk, author of *The Caine Mutiny* and *The Winds of War,* and a World War II Navy officer himself, told an audience emerging from the disheartening experience of Vietnam, that in a society riven by social and political turmoil, their job was "Not to solve the great ongoing problems of social stress, nor to despair at the immensity and complexity of these problems outside our country and inside, but to stand and to serve. To improvise, to make do with what we have; to serve in still another kind of revolutionary warfare . . . and with this service, to give freedom one more chance for one more generation."[27]

Notes

[1] The President's unique oath is found in the final paragraph of Article II, Section 1, of the Constitution.

[2] Commission document.

[3] Continental Congress, 1777, Printed Commission Form. George Washington Papers at the Library of Congress, 1741–1799: Series 4, General Correspondence, 1697–1799, April 30, 1777, to January 24, 1778, image 676, available at <http://memory.loc.gov/ammem/gwhtml/gwhome.html>.

[4] Washington's 1753 commission in the Virginia Regiment shows *ability, conduct*, and *fidelity* as the relevant virtues. *The Writings of George Washington: Being His Correspondence, Addresses, Messages, and Other Papers Official and Private, Selected and Published from the General Manuscripts; with A Life of the Author, Notes*

and Illustrations, Jared Sparks, ed., 12 volumes, vol. II (Boston: American Stationer's Company, 1838), available at <http://babel.hathitrust.org/cgi/imgsrv/download/pdf?id=uc1.c2741841;orient=0;size=100;seq=461;num=429;attachment=0>.

[5] Quoted in Department of the Navy, Chief of Naval Operations [Admiral Gary Roughead], Memorandum for All Prospective Commanding Officers, Subject: The Charge of Command (June 9, 2011). Emphasis in original. Available at <www.public.navy.mil/bupers-npc/officer/Detailing/Documents/CNO%20Charge%20of%20Command%20Letter%20-%20Signed.pdf>.

[6] U.S. Code, Title 10, "Armed Forces," Sections 5947, 3583, and 85831, "Requirement for Exemplary Conduct," as quoted in Cornell University Law School, Legal Information Institute, available at <www.law.cornell.edu/uscode/text/10>.

[7] Richard Swain, *The Obligations of Military Professionalism: Service Unsullied by Partisanship* (Washington, DC: Institute for National Security Ethics and Leadership, December 2010), 7, available at <www.mca-marines.org/files/obligations%20of%20military%20professionalism.pdf>.

[8] *The Armed Forces Officer* (Washington, DC: U.S. Government Printing Office, 1950), 1. Marshall abandoned these words in his final edition of the book in 1975. Times were hard in the wake of the war in Vietnam, and Marshall was evidently distressed. He wrote in his introductory remarks:

Love of country is still the only possible refuge for intelligent American men and women in service; it is their sword and shield and the emblem of their advance. Everything that enters into the making of truly superior military officers would qualify them to live more generously and rewardingly in any other company. That, essentially, is what this manual has to say. Hardly a new and radically different treatment of the subject, it is at least as old as the American Dream.

"Introduction," *The Armed Forces Officer* (Washington, DC: Armed Forces Information Service, 1975), iii.

[9] *The Armed Forces Officer* (1950), 3.

[10] U.S. Code, Title 5, "Government Organization and Employees," Part III, "Employees," Subpart B, "Employment and Retention," Chapter 33, "Examination, Selection and Placement," Subchapter II. Oath of office, Section 3331, "Oath of Office," as quoted in Cornell University Law School, Legal Information Institute, available at <https://www.law.cornell.edu/uscode/text/5/3331>.

[11] There are two exemplary studies of the oath of office by commissioned officers: Kenneth Keskel, "The Oath of Office: A Historical Guide to Moral Leadership," *Air and Space Power Journal* (Winter 2002), available at <www.airpower.maxwell.af.mil/airchronicles/apj/apj02/win02/keskel.html>; and Thomas H. Reese, "An Officer's Oath," *Military Law Review* (July 1964), 1-41, available at <www.loc.gov/rr/frd/Military_Law/Military_Law_Review/pdf-files/276073~1.pdf>.

[12] *The Armed Forces Officer* (1950), 158.

[13] Sir John Winthrop Hackett, *The Profession of Arms* (New York: Macmillan, 1983), 202.

[14] In re Grimley, 137 U.S. 147, 153 (1890), cited as precedent by Chief Justice William Rehnquist in Parker, Warden, et al. v. Levy, 417 U.S. 733 (1974), available at <https://supreme.justia.com/cases/federal/us/417/733/case.html>.

[15] Hackett, *The Profession of Arms*, 220. The quotation is from an essay titled "Leadership," which was not part of the original lectures.

[16] *See Manual for Courts Martial*, Article 90, "Assaulting or willfully disobeying superior commissioned officer," and compare it to Article 91, "Insubordinate conduct toward warrant officer, noncommissioned officer or petty officer," available at <www.apd.army.mil/pdffiles/mcm.pdf>.

[17] U.S. Department of the Navy, *Marine Corps Manual* with Changes 1-3, 1980; Change 1, 1982; Change 2, 1984; Change 3, 1996, Paragraph 1100, 1–21, 1–22, available at <www.marines.mil/Portals/59/Publications/MARINE%20CORPS%20MANUAL%20W%20CH%201-3.pdf>.

[18] *The Armed Forces Officer* (1950), 2.

[19] Ibid.

[20] Army Doctrine Reference Publication 6-22, *Army Leadership* (Washington, DC: Headquarters Department of the Army, August 2012), 2-1, available at <http://armypubs.army.mil/doctrine/DR_pubs/dr_a/pdf/adrp6_22_new.pdf>.

[21] William T. Sherman, "Address to the Officers and Soldiers Composing the School of Application at Fort Leavenworth, Kansas, October 25, 1882," 8, available at the digital collection of the Archives, Combined Arms Research Library, U.S. Army Command and General Staff College, Fort Leavenworth, Kansas, <http://cgsc.contentdm.oclc.org/cdm/ref/collection/p4013coll4/id/352>.

[22] William Pfaff, "The Honorable Absurdity of the Soldier's Role," *International Herald Tribune On Line*, March 25, 2003, 8. Available at <www.iht.com/ihtsearch.php?id=90840&owner=(IHT)&date=20030326163223>. Reprinted without title, dated March 20, in William Pfaff, *Fear, Anger and Failure: A Chronicle of the Bush Administration's War Against Terror from the Attacks of September 11, 2001 to Defeat in Baghdad* (New York: Algora Publishing, 2004), 169-171. (Punctuation varies in minor instances between versions.)

[23] There is a paradox here, for the whole purpose of tactics at the small unit level and the massive investment in sophisticated hardware at the national level is intended to drive the risk of losses as close to zero as possible. The argument is that at some point this creates a new understanding of the military professional and changes fundamentally the ethical climate in which he or she lives.

[24] Les Brownlee and Peter Schoomaker, "Serving a Nation at War: A Campaign Quality Army with Joint and Expeditionary Capabilities," *Parameters* (Summer 2004), 12–13, available at <http://strategicstudiesinstitute.army.mil/pubs/parameters/Articles/04summer/schoomak.pdf>. The same point was argued by Michael Ignatieff, *The Warrior's Honor: Ethnic War and the Modern Conscience* (New York: Henry Holt, Owl Books, 1998).

[25] James N. Mattis, "Ethical Challenges in Contemporary Conflict: The Afghanistan and Iraq Cases," The 2004 William C. Stutt Ethics Lecture sponsored by the Center for the Study of Professional Military Ethics, U.S. Naval Academy, Annapolis, MD, November 2004, 10.

[26] Ibid., 11–12.

[27] Herman Wouk, Spruance Lecture, "The Naval Officer in an Age of Revolution," *The Naval War College Review* 25, no. 4 (March–April 1973), 10–11, available at <www.usnwc.edu/Publications/Naval-War-College-Review/ArchivedIssues/1970s/1973-March-April.aspx>.

CHAPTER TWO

The Profession of Arms

Humans fight as individuals and as groups. Some fight primarily for money, some for love of fighting, and some for lack of alternative opportunities. Others fight for love of country and civic duty. As noted by General Sir John Hackett, "From the beginning of . . . recorded history physical force, or the threat of it, has always been freely applied to the resolution of social problems."[1] Human societies—from tribes and city-states to empires, organized religions, and nation-states—have regularly established and relied on groups of specialists who, willingly or unwillingly, assumed the burden of fighting, killing, and dying for the larger group. Whatever the formal name or title given to these groups, theirs is the profession of arms.

It is a basic premise of civilized societies, especially democratic ones, that the military serves the state (and by extension, the people), not the other way around. The profession of arms exists to serve the larger community, to help accomplish its purposes and objectives, and to protect its way of life. As Samuel Huntington put it in *The Soldier and the State*: "The justification for the maintenance and employment of military force [or military *forces*, for that matter] is in the political ends of the state."[2] In wartime or in peacetime, at home or abroad, the Armed Forces serve the larger society and perform the tasks their government assigns them.

In his classic study *The Profession of Arms*, General Hackett stated, "The function of the profession of arms is the ordered application of force in the resolution of a social or political problem."[3] The essential task of its members is to fight, individually and collectively; of its officers, to direct and lead those who apply the instruments of destruction to achieve assigned ends. With rare exceptions, a society's government identifies the problems to be resolved with force, and it then turns to and relies on the professionals to handle the always difficult, usually dangerous, often bloody details in a manner acceptable to the citizens and supportive of their goals.

The most basic task of the profession of arms is the armed defense of the society, its territory, population, and vital interests. In its most elemental sense, the profession of arms is all about fighting and all about *war*. As the 19th-century Prussian strategist and student of war Carl von Clausewitz observed, "For as long as they practice this activity, soldiers will think of themselves as members of a kind of guild, in whose regulations, laws, and customs the spirit of war is given pride of place."[4] The defining mission of the Armed Forces is the preparation for and the conduct of war, which includes securing the military victory until peace is restored politically. It is the warfighting mission that determines how forces are organized, equipped, and trained.

Whatever its particular forms, this unique and specialized service to the Nation gives the military profession its own nature and distinctive status. Because those responsibilities include the potentially wholesale taking and losing of life, the military profession stands alone, in its own eyes and in the eyes of those it serves. Its members must always be conscious of their commitment: to be prepared to give that "last full measure of devotion."[5] They serve at frequent cost to their convenience, comfort, family stability, and often their limbs and lives. It is ultimately because of their willingness to endure hardship and risk life and limb on behalf of the Nation, not the willingness to kill and destroy in the Nation's name, that members of all the Armed Forces enjoy the respect and gratitude of the American people. Theirs is a higher loyalty and purpose, or rather a hierarchy of loyalties, which puts nation above service, service above comrades in arms, and comrades above self. Soldiers serve the Nation; they fight and die for each other.

The commitment to the Nation is a two-way street between the individual military member and the larger society. Society invests much—its safety and security, its hopes and ideals, much of its treasure, and the best of its men and women—in the Armed Forces. For the member of the profession of arms, fulfilling society's demands and expectations means investing one's best as a professional and as a person. As General Hackett observed, "Service under arms has been seen at times and in some places as a calling resembling that of the priesthood in its dedication."[6]

Like the priesthood, the profession of arms is a *vocation,* a higher calling, to serve others, to sacrifice self, to be about something larger than one's own ambitions and desires, something grander than one's own contributions and even one's own life. This is a recurring and central theme in discourses on the profession of arms. Reflecting on "General George C. Marshall and the Development of a Professional Military Ethic," Josiah Bunting III noted that the "ethical leadership of George Marshall provided many lessons[s] including: an officer never is to take the counsel of his ambition."[7] At the dedication of the U.S. Army War College, Secretary of War Elihu Root told the assembled audience and, by extension, all military members: "Remember always that the highest duty of a soldier is self-abnegation. Campaigns have been lost for no other cause than the lack of that essential quality."[8]

This hierarchy of loyalties has several formulations in the United States Armed Forces. In the Air Force, it is "service before self." In the Navy, it is "ship-shipmate-self." The Army defines the value of loyalty as a hierarchy of responsibilities to the Constitution, service, unit, and other Soldiers. The basic idea is that there is always something larger, something more important than the individual. Service in the Armed Forces is not primarily about self, but rather about others—fellow citizens and fellow military members. In Huntington's words, "The military ethic is basically corporative in spirit."[9]

The loyalty to fellow military members has its roots and its rationale in the ultimate activity of the Armed Forces—combat and war. What Lieutenant General Harold G. Moore, USA (Ret.) and Joseph L. Galloway wrote in the prologue to their memorable book about Vietnam could have been said by soldiers of any nation about any war: "We discovered in that depressing, hellish place, where death was our constant companion, that we loved each other. We killed for each other, we died for each other, and we wept for each other."[10] The classic statement of this perennial and honorable theme is in Shakespeare's *Henry V*:

We few, we happy few, we band of brothers;
For he today that sheds his blood with me
Shall be my brother.

Given the stakes, it is no wonder that the profession of arms invokes and requires, in the words of the U.S. military officer's commission, "special trust and confidence."

"The modern officer corps is a professional body and the modern military officer is a professional man."[11] So wrote Huntington in 1957, in the first sentence of chapter 1 of *The Soldier and the State.* Historians would dispute that the status was recent, or even unassumed, in 1957. Some parts of this sentence, such as the masculine noun and its restriction to the officer corps, are now out of date. But Huntington's basic thesis was that the military belonged in the ranks of the classic professions, including the clergy, medicine, and law. The military possessed what Huntington took to represent the "distinguishing characteristics of a profession as a special type of vocation . . . *expertise, responsibility,* and *corporateness* [emphasis added]."[12] Experience has shown the importance of a fourth characteristic, a professional ethic and an *ethos.*

For Huntington, as well as other authors, *profession* is not a term to be thrown about loosely. The concept of "a profession" is an abstract, inductive, descriptive device adopted by 19th- and 20th-century social scientists to examine similarities and differences among characteristics present in particular kinds of human organizations for work—particularly medicine, law, and clergy. Experts disagree somewhat on the particulars of those characteristics, and their relative importance, but tend to agree on this point: "A profession is a peculiar type of functional group with highly specialized characteristics."[13] The nature and forms of professions evolved significantly in the 20th century, and it is safe to say that the structure and organization of the medical profession, the paradigmatic case, has changed a good deal since Huntington wrote in the mid-1950s.

Huntington's basic argument—that the modern military is a profession—is widely accepted today, certainly in the United States.[14] The concern now is not to *prove* that the military is a profession, but rather to inspire men and women in uniform to reflect the expected characteristics of professionals in their day-to-day activities: to hold themselves and others to uniformly high standards of performance and conduct, lest they lose the discretion in performance that is the acknowledgment of professional status. On his first day as Chairman of the Joint Chiefs of Staff, General Martin E. Dempsey wrote a letter to the Joint Force in which he identified his key themes as Chairman, one of which was: "We must renew our commitment to the Profession of Arms. We're not a profession simply because we say we're a profession. We must continue to learn, to understand, and to promote the knowledge, skills, attributes, and behaviors that define us as a profession."[15] For General Dempsey and for others, it is not in the *saying* but in the *doing* that the heart of a profession lies.

Influenced by Huntington, General Hackett wrote that the military occupation

> *has evolved into a profession, not only in the wider sense of what is professed, but in the narrower sense of an occupation with a distinguishable corpus of specific technical knowledge and doctrine, a more or less exclusive group coherence, a complex of institutions peculiar to itself, an educational pattern adapted to its own specific needs, a career structure of its own and a distinct place in the society which has brought it forth.*[16]

This chapter describes four elements that are widely accepted as characteristic to any profession: special expertise, a collective and individual responsibility to serve society, a sense of "corporateness," and a professional ethic and ethos.

Expertise

A distinguishing characteristic of any profession is authority for discretionary application of a unique knowledge, based on society's implicit trust that members will apply their particular skills reliably, effectively, honorably, and efficiently. Thus, a profession is an identifiable body of practitioners granted authority (by the larger society) for discretionary practice of a unique and necessary skill.

A profession has a body of *expertise*, built over time on a base of practical experience, which yields fundamental principles and abstract knowledge; which normally must be mastered through specialized education; which is intensive, extensive, and continuing; and which can then be applied to the solution of specific, practical problems. "Professional knowledge . . . is intellectual in nature and capable of preservation in writing. Professional knowledge has a history, and some knowledge of that history is essential to professional competence."[17]

The body of specialized knowledge changes over time, as various factors evolve or new ones appear. One responsibility of a profession and of its individual members is to acquire and apply this new information, integrating or synthesizing it into the existing body of knowledge. This is done through formal education, in professional schools, and through individual and collaborative experiential learning "on the job." Individual professionals share experiences, insights, and knowledge, engage in continuous learning, and serve as faculty or instructors in various professional schools and courses. Continuing self-development is one of the hallmarks of a profession and its individual members.

In describing the expertise of the profession of arms, Huntington used political scientist Harold Lasswell's phrase "the management of violence," which he went on to say involves "(1) the organization, equipping, and training of [the] force; (2) the planning of its activities; and (3) the direction of its operation in and out of combat."[18] Many will recognize in the first category the functions that Title 10 of the U.S. Code assigns to the three military departments[19] (Army, Air Force, and Navy).[20] Much of the second and third types of work is done, in the United States, by the Combatant Commanders and the Chairman of the Joint Chiefs of Staff under guidance and direction from the Commander-in-Chief and the Secretary of Defense.

The management of violence draws on a body of knowledge developed over centuries, through organized reflection on historical and personal experiences; from this reflection come abstract principles, which have been honed, transmitted, and advanced in professional military education institutions, so that military professionals can apply them to the solution of practical military problems. In book two of *On War,* Clausewitz explains how military theory grew out of the reflections of individual warriors on their own personal experiences, especially in war:

> *As these reflections grew more numerous and history more sophisticated, an urgent need arose for military principles and rules whereby the controversies that are so normal in military*

> *history—the debate between conflicting opinions—could be brought to some sort of resolution. . . . Efforts were therefore made to equip the conduct of war with principles, rules, or even systems.*[21]

The traditional notion was that this specialized knowledge in the management of violence was to be applied to "fight and win the nation's wars." However, this traditional notion does not exhaust the variety of tasks societies give their organized and uniformed fighters. Because they are disciplined and armed organizations, with a wide range of skills and capabilities, military forces are called upon frequently to perform other important missions in service to the state, such as maintaining civil order at home and abroad and providing disparate forms of civil relief in times of crisis or disaster. It is important not to think that the primary mission for which the Armed Forces are organized, trained, and equipped is the *only* mission society may legitimately give them.

Society may change the terms of the services that it expects, or even demands, a particular profession will provide. Accordingly, a desire on the part of citizens to change the definition of the services they expect can lead the profession to expand the range of services it has traditionally provided. In the United States, the Army, in particular, has been used at various times to perform internal development, to promote exploration, to maintain order, to enforce Federal law, and even to run Civilian Conservation Corps camps during the Great Depression of the 1930s. The U.S. Coast Guard's principal roles involve maintaining the security and safety of the Nation's ports and waterways and enforcing Federal laws and treaties on the high seas. Traditionally, detachments of Marines guard U.S. embassies abroad, and Air Force and Navy lift assets and technical units are regularly pressed into service providing transportation for relief supplies in disasters at home and abroad. The organizational and planning skills of Armed Forces officers are often transferable to nontraditional assignments, and no less valuable than their material contributions. Sir David Richards, a former Chief of Defense Staff in Great Britain, writes that: "The armed forces' great strength lies in our capacity to analyse a problem, plan a solution and then implement it under pressure."[22] The U.S. Armed Forces are expected to bring great skill and enthusiasm to all assignments.

Service to Society

A profession has a responsibility to provide a useful, even critical, service to the larger society. In exchange for the service that a profession provides, the society grants to members of that profession certain privileges, prerogatives, and powers that it does not extend to the rest of its citizens.

The American people have granted the Armed Forces: custody of nuclear weapons; extraordinary latitude in managing their own affairs, including their own legal code (the Uniform Code of Military Justice[23]); the Federal courts' customary reluctance to interfere with the chain of command's management of good order and discipline; a high degree of discretion in the use of lethal force to accomplish assigned missions; and a set of benefits beyond the reach, or claim, of most citizens. The traditional deference to military management of military affairs is not absolute. Society, especially in a democratic political system, always reserves the right to intervene when it thinks that military values and practices should change to conform to public norms. Article I of the Constitution vests in the Congress the power and the authority "To make Rules for the Government and Regulation of the land and naval Forces," and Article II vests chief command in the President of the United States.

Others outside the profession may claim equivalent or superior expertise, and challenge the "monopoly" of relevant knowledge that the recognized military profession has traditionally claimed and enjoyed. This can lead to jurisdictional disputes over who is a professional and who may legitimately provide certain services to the public.

In the second half of the 20th century, with the emergence of nuclear weapons, the purposes on which the military's specialized knowledge focused were transformed, to include something that had never been even imagined by previous generations of military professionals: *nuclear deterrence.* Nuclear deterrence appeared to many soldiers as a condition analogous to traditional uses of military force, but it was qualitatively different because of the magnitude and imminence of continuous catastrophic threat. The addition of this new and critical concern for the traditional body of specialized knowledge led to a "jurisdictional dispute" with experts outside the uniformed community and to the emergence of civilian nuclear strategists, not military officers, as the dominant intellectual force in the development and evolution of nuclear strategy. The results were paradigm-changing. In 1946, strategic analyst Bernard Brodie wrote, "Thus far the chief purpose of our military establishment has been to win wars. From now on its chief purpose must be to avert them."[24]

No military professionals had any experience, or even any theoretical background, in using nuclear weapons to *deter* war. So, enter the economists, game theorists, systems analysts, operations research specialists, historians, and political scientists, who, taken together, claimed to have more expertise relevant to deterring nuclear war than the uniformed military. Thus, the classic texts on nuclear strategy were written by civilians, and not by uniformed military professionals in whose hands execution of nuclear war largely remained, albeit with significant close oversight. At the same time, military practitioners were forced to rethink the use of conventional military forces within the context of a potentially nuclearized strategy. Moreover, in a world threatened by nuclear oblivion, the discretion of military practitioners everywhere was significantly curtailed. Global communications permitted the President and his civilian aides to become involved in military execution at levels never dreamed of by President Lincoln in the War Department telegraph office. Professional autonomy was significantly reduced.

The last decades of the 20th century saw the rise of defense consulting firms and nonprofits, concentrated in the Washington, D.C. area, which now compete aggressively with statutory advisors in providing advice on the full range of military-strategic subjects to both executive and legislative branches. Numerous retired military officers have found a lucrative base for continued involvement as rivals of their appointed successors, or substitutes for required personnel beyond Congressional manpower ceilings within the defense structure. Yet another contemporary example of such a jurisdictional shift can be seen since 2001 in the unprecedented use of civilians working for private security companies to do many tasks performed since the 18th century largely by uniformed members of the Armed Forces. In short, the assumed monopoly of uniformed professionals over the practice of supposedly unique military skills has become contested throughout the field of military practice.

Corporateness

A profession has a sense of what Huntington called, somewhat awkwardly, *corporateness*, which he defined as "a sense of organic unity and consciousness of themselves as a group apart."[25] There are at least two important dimensions of this corporateness: a shared identity, and the wish to exert control over membership in the profession. The shared identity comes from the culture and ethos of a profession.[26] It reflects a sense of common endeavor and can be manifested in the adoption of distinctive titles and/or distinctive attire, and reciprocal recognition of members. The titles and attire are visible manifestations of a deeper, invisible identity shared by the members of the profession.

In the Armed Forces, the most visible manifestation of this shared identity is the uniform. "The uniform regulations of the Navy, for instance, point out that 'uniforms are distinctive visible evidence of the authority and responsibility vested in their wearer by the United States.'"[27] More broadly, this identity as members of the Armed Forces of the United States (or any other country) is shared among those "who wear the cloth of the nation." U.S. law generally prohibits wearing of the uniform other than by members of the Armed Forces of the United States.[28]

The more practical aspect of corporateness is that the members of the military profession have significant influence over the criteria for entrance into the profession.[29] They exercise this influence by setting and enforcing standards for practice, standards that are made public and must be publicly defensible. In the United States, or for that matter in most (maybe all) countries, no man or woman can declare himself or herself to be a Soldier, Marine, Sailor, Airman, or Coastguardsman.[30] Rather, in the United States, persons who aspire to that status must apply to join. The individual Service then screens the candidates according to public standards set by the Congress and Department of Defense, and accepts some applicants conditionally. These applicants are sent to one form or another of initial military training. Upon successful completion of that training, the Service then accepts those individuals officially into its ranks. "Certification and testing to become a full-fledged professional member of the Armed Forces are achieved upon completion of specific [initial military training requirements] where one earns the title of Soldier, Marine, Sailor, Airman, or Coastguardsman." Indeed, a fundamental purpose of any initial military training is to help transform a civilian into a military professional.

Beyond initial acceptance into the ranks of the profession of arms, it is the profession, through the evaluation by its leadership at various levels and not solely the desire of the individual military member, that determines who remains and who advances in the profession—and who must leave, again according to publicly stated standards promulgated under authority of law.

Advancing in the profession is, of course, reflected in promotions to successively higher ranks and positions of increasing authority. This process reflects an older notion of stages in a career, one that goes back to the medieval guilds with apprentice, journeyman, and master levels. In the words of a former Sergeant Major of the Army, "Just as other professions have entry level or apprentice, mid-level or journeyman, and senior or expert levels within their professions, we have levels of competence within our Army."[31]

In some cases, an individual military member may wish to stay in uniform, but the professional leadership of the Service has determined, through one mechanism or another, and for one reason or another, that he or she has not measured up to the standards of the profession or is believed to possess less potential than others for future success, and thus will be discharged or permitted to retire. In this way, commissioned officers are central actors in setting and enforcing the standards for membership or advancement in the profession of arms.

Ethics and Ethos

Professional status is reflected most dramatically in a body of *professional ethics* and a *professional ethos,* which are related but not identical. Professional ethics are the moral standards to which the profession is committed and held. Much of the professional ethic is spelled out in official documents, such as Title 10 of the U.S. Code, the Uniform Code of Military Justice, and the Code of Conduct for members of the Armed Forces of the United States.[32] In contrast, a professional *ethos* is the collective and *internal* sense of what each member must *be* as a member of the profession. It is felt more than known. In ancient Greek, *ethos* meant what is customary. Customs and expected behaviors lend much of the flavor that any profession is said to possess. In this sense, the *ethos,* which includes the tribal wisdom and oral tradition handed on from one generation to the next, is the standard-bearer of the profession.[33]

An ethos is more intangible than a professional ethic, though its importance is central to the notion of a profession, and more especially to a *professional identity.* An ethos is more about what it means to *be* a member of that profession than it is about what members of the profession *do.* One *must do* certain things because one *is* a member of a certain profession, and one *must not do* certain other things, also because one *is* a member of the profession. In many important ways, the ethos is the defining characteristic of any profession.

The professional military ethos includes much that is written but not official or authoritative, such as General of the Army Douglas MacArthur's famous "Duty, Honor, Country" speech at West Point,[34] the Marine Corps' motto *Semper Fidelis,* the Coast Guard's *Semper Paratus*, and the spirit of each military Service's core values. It also includes much that is not written down or published at all, much that is intangible but nonetheless central to the identity that makes a Soldier, Marine, Sailor, Airman, or Coastguardsman. Service ethos is the foundation of *esprit de corps*, the "sense of unity and of fraternity in its routine existence which expresses itself as the force of cohesion in the hour when all ranks are confronted by common danger."[35]

Because, in the commission, the President of the United States reposes "special trust and confidence in the patriotism, valor, fidelity, and abilities" of the named individual, officers have particular and weighty responsibilities as custodians of the profession of arms.

A useful framework for professional military ethics and its ethos has three parts: the Individual in the Profession, the Profession at Work, and the Profession and Society. The next several chapters elaborate these three categories of professional military ethics, in particular how they apply to officers: the profession of arms, the ethical use of force, leadership, command, civilian control of the military, and the military's and society's values.

Notes

[1] Sir John Winthrop Hackett, *The Profession of Arms* (New York: Macmillan, 1983), 9.

[2] Samuel P. Huntington, *The Soldier and the State: The Theory and Practice of Civil-Military Relations* (Cambridge: Belknap Press, 1985), 65.

[3] Hackett, 9.

[4] Carl von Clausewitz, *On War*, ed. and trans. Michael Howard and Peter Paret (Princeton: Princeton University Press, 1976), 187.

[5] Abraham Lincoln, "Address Delivered at the Dedication of the Cemetery at Gettysburg, 19 November 1863," in *Abraham Lincoln: Great Speeches* (Mineola, NY: Dover Publications, 1991), 104.

[6] Hackett, 9.

[7] Josiah Bunting III, "General George C. Marshall and the Development of a Professional Military Ethic," *Footnotes* 16, no. 4 (June 2011), 4.

[8] Elihu Root, *The Military and Colonial Policy of the United States* (Cambridge: Harvard University Press, 1916), 128-129.

[9] Huntington, 64.

[10] Harold G. Moore and Joseph L. Galloway, *We Were Soldiers Once . . . and Young* (New York: Harper Perennial, 1993), xviii.

[11] Huntington, 7.

[12] Ibid., 8.

[13] Ibid., 7.

[14] The Army has taken the position that there is an "Army Profession," encompassing all members of the Army institution, distinct from the "Army Profession of Arms" that is made up only of uniformed members. The Army defines this "Army Profession" as a "unique vocation of experts certified in the design, generation, support, and ethical application of landpower, serving under civilian authority and entrusted to defend the Constitution and the rights and interests of the American people." The Army has issued a publication explaining the various canons of membership. See Army Doctrine Reference Publication (ADRP) 1, *The Army Profession* (Washington, DC: Headquarters Department of the Army, June 2015), "Preface," v, and paragraphs 1–10, 1–2.

[15] Martin E. Dempsey, "General Dempsey's Letter to the Joint Force," October 1, 2011, available at <www.dodlive.mil/index.php/2011/10/general-dempseys-letter-to-the-joint-force/>.

[16] Hackett, 9.

[17] Huntington, 8.

[18] Ibid., 11.

[19] See U.S. Code, Title 10, Sections 5013 (b)—Secretary of the Navy; 8013 (b)—Secretary of the Air Force; and 3013 (b)—Secretary of the Army.

[20] The Department of the Navy includes two military Services: the Navy and Marine Corps.

[21] Clausewitz, 134.

[22] General Sir David Richards, *Taking Command: The Autobiography* (London: Headline Publishing Group, 2014), an e-book on Kindle, position 4993 of 6468 (chapter 16). For a related article about the general presence of a high level of managerial skills in the officer corps, see Adam Davidson, "Rebuilding the Middle Class the Army Way," *The New York Times Sunday Magazine* (December 15, 2015), MM16, available at <www.nytimes.com/2015/12/20/magazine/rebuilding-the-middle-class-the-army-way.html?rref=collection/sectioncollection/magazine&action=click&contentCollection=magazine®ion=rank&module=package&version=highlights&contentPlacement=8&pgtype=sectionfront&_r=1>.

[23] U.S. Code, Title 10, Chapter 47.

[24] Bernard Brodie, *The Absolute Weapon* (New York: Harcourt Brace, 1946), 78.

[25] Huntington, 10.

[26] See, for example, Ann E. Rondeau, "Identity in the Profession of Arms," *Joint Force Quarterly* 62 (3rd Quarter 2011), 10–13.

[27] Navy Personnel Command, "United States Navy Uniform Regulations," available at <www.public.navy.mil/bupers-npc/support/uniforms/uniformregulations/Pages/default.aspx>, 32.

[28] U.S. Code, Title 10, Section 771—Unauthorized Wearing Prohibited, states: "Except as otherwise provided by law, no person except a member of the Army, Navy, Air Force, or Marine Corps, as the case may be, may wear—(1) the uniform, or a distinctive part of the uniform, of the Army, Navy, Air Force, or Marine Corps; or (2) a uniform any part of which is similar to a distinctive part of the uniform of the Army, Navy, Air Force, or Marine Corps." The statute goes on to identify a few specified exceptions to the general prohibition.

[29] Congress, of course, retains its Article I powers cited above, and civilian officials in the executive branch retain their Article II powers.

[30] *The Noncommissioned Officer and Petty Officer: Backbone of the Armed Forces* (Washington, DC: NDU Press, 2013), 24.

[31] Former Sergeant Major of the Army Kenneth Preston, as quoted in *The Noncommissioned Officer and Petty Officer*, 24.

[32] Executive Order 10631, "Code of Conduct for members of the Armed Forces of the United States," available at <www.archives.gov/federal-register/codification/executive-order/10631.html>.

[33] Michael Boylan, *Basic Ethics* (Upper Saddle River, NJ: Prentice Hall, 2000), 154.

[34] Douglas MacArthur, "Thayer Award Acceptance Speech," West Point, NY, May 12, 1962, available at <www.americanrhetoric.com/speeches/PDFFiles/Douglas%20MacArthur%20-%20Thayer%20Award%20Address.pdf>.

[35] *The Armed Forces Officer* (Washington, DC: U.S. Government Printing Office, 1950), 164.

CHAPTER THREE

The Officer in the Profession of Arms

Armed Forces officers are the appointed leaders of the uniformed component of an executive department of government. They are viewed as professionals, contingent upon their demonstrated abilities to deliver competent, reliable, discretionary service of a unique and necessary kind. Because they serve in a hierarchy of rank and authority, all Armed Forces officers are simultaneously leaders and followers, bound by their oath and commission to loyal subordination as well as effective direction of others. They are called upon by overlapping demands to display a number of virtues, some inherent in the terms of their commissions; some reflecting values adopted for all members by their respective departments to ensure faithful reliable service; still others of the sort commonly found in all skilled professions to guarantee the excellence and continued relevance of the discretionary service on which are based the claims for authority to practice their unique skills. The Armed Forces officer is expected to synthesize all these virtues into a harmonious whole, and to practice their application self-consciously, until they become second nature.

The first chapter addressed the expectations expressed in the commission for virtues of patriotism, valor, fidelity, and abilities; the nature and requirements of the constitutional oath; and the admonition for disciplined service explicit in both the commission and oath of office. These are the basis of entry into commissioned service, and they are supplemented and undergirded by other expectations and requirements of service. Among the first the new officer will confront are Service values, promulgated in each case under the authority of the respective Service secretary.

Service Core Values and Military Virtues: A Shared Ethic and Ethos

S.L.A. Marshall attributed the commissioning of the first edition of *The Armed Forces Officer* to the conviction of George C. Marshall "that American military officers, of whatever Service, should share common ground ethically and morally."[1] Each of the U.S. Services has a set of institutional core values that aim to describe and define what it means to be a Soldier, a Marine, a Sailor, an Airman, or a Coastguardsman. In the aggregate, they might be said to illustrate George Marshall's conviction.

Each Service expects its members both to exhibit these virtues and to demand them from members who may become lax in their performance. This is what is meant by corporateness in a profession or *esprit de corps* in a military unit. S.L.A. Marshall wrote: "The man who feels the greatest affection for the service in which he bears arms will work most loyally to make his own unit know a rightful pride in its own worth."[2] He argued that the Marine Corps was most faithful to this principle.

The point is not that these virtues or qualities are absent in the civilian world, but rather that they take on a new and profound meaning in the profession of arms. General John Hackett wrote that the military virtues such as courage, fortitude, and loyalty are functionally indispensable for officers, "not just because they are morally

Table. Service Values

U.S. Army	U.S. Navy and Marine Corps	U.S. Air Force	U.S. Coast Guard
Loyalty	Honor	Integrity First	Honor
Duty	Courage	Service Before Self	Respect
Respect	Commitment	Excellence in All We Do	Devotion to Duty
Selfless Service			
Honor			
Integrity			
Personal Courage			

desirable in themselves, but because they contribute to military efficiency."[3] Officers recognize a set of reciprocal expectations binding each to those with whom they serve. Officership, the practice of *being* professional officers and leaders, requires an internalization and self-conscious understanding of a personal obligation to the ethos of the profession and to all those who depend upon the quality of their individual service. Substantive similarities among Service values are obvious; apparent differences can for the most part be understood in terms of traditions and outlooks specific to the individual Services.

Service core values are an integral, indeed central part of initial military training in all five Services. They feature prominently in each Service's presentation of itself to its membership and the public. They represent institutional goals to which all members are expected to aspire in their personal and professional conduct. When internalized and reflected in one's habitual behavior, values become virtues.

A virtue is a "persisting, reliable and characteristic" feature that produces a disposition in an individual to behave in a certain desirable way.[4] Once a virtue is ingrained in a person, he or she should act naturally in accordance with the value it represents. Knowledge of values is not enough. It is *the will to act* in accordance with them that transforms a value into a virtue. The profession of arms demands constant self-awareness, self-reflection, and self-criticism of the times and places where better, more virtuous choices should have been made.

For Aristotle, developing a virtue is a matter of habituation: "Moral goodness . . . is the result of habit."[5] Most drill instructors would not think of themselves as disciples of Aristotle, but in reality they are. It is through repetitive actions that one acquires a virtue. A recruit becomes obedient by obeying the drill instructor, on things large and things small, over and over and over again. An officer becomes virtuous by disciplined and reflective effort to live up to the imperatives of the oath and the commission, the expectations of the Nation, and the obligations of the officer's service to the Nation.

As stated in the previous chapter, an ethos is more about *who you are* than it is about *what you do.* Who you *are* determines what you *do* and *do not do.* The Honor Concept of the Brigade of Midshipmen at the U.S. Naval Academy aptly reflects this basic principle:

- Midshipmen are persons of integrity: They stand for that which is right.
- They tell the truth and ensure that the truth is known. *They do not lie.*
- They embrace fairness in all actions. They ensure that work submitted as their own is their own, and that assistance received from any source is authorized and properly documented. *They do not cheat.*
- They respect the property of others and ensure that others are able to benefit from the use of their own property. *They do not steal.*[6]

The text begins with who Midshipmen *are*—persons of integrity—and it goes on to describe what they do and what they do not do *because* they are persons of integrity.

In the U.S. Marine Corps, the admonitory phrase "Marines don't do that," spoken by one Marine to another, recalls the common standard and is an outward reflection of an inner virtue.[7] It is premised on a common respect for the reputation of the Corps and a shared will to demand that all members uphold its ideals. Marines don't *do* that because they *are* Marines.

Like the Service values, there are overlapping lists of the critical military virtues. General Hackett spoke of the military life demanding human qualities of "fortitude, integrity, self-restraint, personal loyalty to other persons, and the surrender of the advantage of the individual to a common good."[8] In a 2014 Veterans Day speech at Georgetown University, Lieutenant General H.R. McMaster, USA, listed the virtues of honor, duty, courage, and loyalty as the basis of a "warrior ethos," binding members of the profession of arms into a self-conscious community. Paraphrasing Professor Christopher Coker, he stated the warrior ethos "permits servicemen and women to see themselves as part of a community that sustains itself through "sacred trust" and [serves as] a covenant that binds us to one another and to the society we serve."[9]

The word *covenant* is important. Coker distinguishes between covenants and contracts, writing of the former: "First, they are not limited to specific conditions and circumstances; secondly, they tend to be open-ended and long-lasting; and, thirdly, they rarely involve individual advantage." Contracts depend on enforceability. "Moral covenants are different. We adhere to rules because of conscience. We obey the dictates of our hearts. We don't wish to dishonour ourselves in the eyes of our moral equals—our friends—and thereby dishonour the unit, the flag or the great tradition."[10] *Marines don't do that!* One is reminded of the covenants between God and man.

Chapter 2 described four characteristics common to all professions: expertise, responsibility, corporateness, and a shared ethic and ethos. These are all woven through the U.S. profession of arms. The Service core values themselves constitute much of a *shared professional ethic*. Adherence to the oath and commission and the obligation to deliver reliable, effective, honorable, and efficient service require an *expertise* guaranteed only by individual dedication to life-long practice and learning. Responsibility as an expectation of right action also is inherent in both the oath and commission. It will be addressed more fully later, in chapters focused on leadership and command. For now, it may be said simply that the virtue of *responsibility*, as a desideratum of professional service, involves clarity of motivation. It demands that officers develop the courage to act—to decide, to direct, to follow through—*and* to accept accountability for the consequences of the outcomes of their decisions and actions. Responsibility involves recognition of an institutional anticipation of right conduct by officers under all circumstances.

Finally, *corporateness* involves acknowledgement of the shared responsibility to maintain and display mutual respect for fellow members of the profession, regardless of rank or Service or specialty, ethnic origin or gender. It involves no less than granting others recognition of kinship and presumption of good intention, unless there is evidence to the contrary. Corporateness requires individual insistence on maintenance of high standards by all members, and adds an obligation for the individual to participate in corporate or institutional learning by sharing his or her own experiences and insights, taking part in professional discourse to explore new problems or find new solutions to older ones under new conditions, and observing continuously what others do and learning from their experiences. All these are reflected in the military life, in aspiration if not entirely full achievement.

What is significant, then, about the characteristics of a profession, is how much they are reflected within traditional military virtues and way of life, an already existing ethic and ethos. Recognition of military service as a profession is achieved, not by the Armed Forces endeavoring to become something other than what they are, so much as by their members living up to the traditional and inherent virtues of military service as they long have been, not because they strive to be recognized as something different, but to live up to their very nature.

Character and Character Development

Discussion of virtues leads naturally to discussion of character and character development, which is the manifestation of the ethic and ethos of the profession of arms. In 1941, General George Marshall told the first graduates of the Army Officer Candidate School that what would enable them to lead men in battle was less their tactical and technical competence, both of which were necessary, or their reputation for courage, but the "previous reputation you have established for fairness, for that high-minded patriotic purpose, that quality of unswerving determination to carry through any military task assigned to you."[11]

Character consists of the set of ingrained virtues, a complex of value-laden dispositions to act reliably, in a particular way, based on one's understanding of the circumstances. James Davidson Hunter of the University of Virginia captures the essence of character succinctly:

> *What, then, can be said about this thing we call character? The most basic element of character is* moral discipline. *Its most essential feature is the inner capacity for restraint—an ability to inhibit oneself in one's passions, desires, and habits within the boundaries of a moral order. Moral discipline, in many respects, is the capacity to say "no"; its function, to inhibit and constrain personal appetites on behalf of a greater good. The idea of a greater good points to a second element,* moral attachment. *Character, in short, is defined not just negatively but positively as well. It reflects the affirmation of our commitments to a larger community, the embrace of an ideal that attracts us, draws us, animates us, inspires us.*[12]

Though he makes no reference here to the military, Hunter in effect points to the link between the character of an individual military member and the ethos of the profession of arms.

There are no "time outs" from exemplary character for officers. As General Marshall told the Officer Candidate School graduates:

> *Never for an instant can you divest yourselves of the fact that you are officers. On the athletic field, at the club, in civilian clothes, or even at home on leave, the fact that you are a commissioned officer in the Army imposes a constant obligation to higher standards than might ordinarily seem normal or necessary for your personal guidance. A small dereliction becomes conspicuous, at times notorious, purely by reason of the fact that the individual concerned is a commissioned officer.*[13]

Ultimately, it is faithfulness to self-understanding that is the basis of an officer's individual integrity and sense of duty—the determination to be, in the words of General Douglas MacArthur, "What you ought to be. What you can be. What you must be."[14] How do institutional values come to be reflected in individual virtues? They do so, borrowing an old line, in the way an aspiring musician gets to Carnegie Hall: by "practice, practice, practice." Effective, reliable, honorable, and efficient service is the officer's obligation to and the expectation of the Nation. Effective service is produced by repetitive training to standard. Exercise of the virtues is intended to produce behavioral habits that result in moral-ethical reliability, guaranteeing honorable and efficient service.

Character development involves training the will as well as the intellect. It is no accident that the U.S. Service academies have long invested considerable time, talent, and resources in *character development* programs as key elements in their overall effort to form young civilians into future military officers who will be men and women of character. More than the other uniformed Services, the Marine Corps grounds its institution explicitly on the character transformation it produces through intensive indoctrination of officer and enlisted aspirants. In the foreword to Marine Corps Warfighting Publication 6-11, *Leading Marines*, General Carl Mundy called character transformation "the education of the heart and of the mind."[15] As the manual itself relates, "Self-image is at the heart of the Marine Corps—a complex set of ideals, beliefs, and standards that define our Corps. Our selfless dedication to and elevation of the institution over self is uncommon elsewhere."[16]

Contemporary practitioners of character development generally focus their efforts on children and young adults and not on mature men and women whose character has been formed years before and for whom it is often too late to form or develop their character anew.[17] Mature adults can be reminded of the values, qualities, characteristics, and virtues that constitute individual or institutional norms or expectations, but whether they choose to act in accordance with the tenets of character, or contrary to them, remains a function of free will—and disposition. As one distinguished retired senior officer stated about his peers who commit various offenses, "They know it is wrong, but they do it anyway." Their weakness is one of will, not understanding. Only focused individual effort, reflection, self-assessment, and a conscious effort to do better will lead formed adults to modify their behavior.

Moral-ethical reliability is vital, because, as General John Vessey—a first sergeant at Anzio in World War II who rose to be Chairman of the Joint Chiefs of Staff in the 1980s—put it bluntly in the 1984 commencement address at the Naval War College:

> *There will not be any tribunal to judge your actions at the height of battle; there are only the hopes of the citizenry who are relying upon your integrity and skill. They may well criticize you later amid the relative calm of victory or defeat. But there is a critical moment in crisis or battle when those you lead and the citizens of the nation can only trust that you are doing what is right. And you develop that concept through integrity.*[18]

Leaders and Followers[19]

While officers exist largely to exercise authority over subordinates, it is also a defining characteristic of military service in the United States that every uniformed member of the Armed Forces is responsible and accountable to a superior, normally more than one. Armed Forces officers exist in a professional hierarchy. Professional loyalty between leaders and followers must be assumed, as must fundamental integrity.

It is in the superior's interest to create an environment in which honest communication is the norm, in which discourse is forthright, and mutual expectations for candor are clear to all. Superior officers have an interest in honest communication, because they very often depend upon the perceptions of subordinates to form their own understanding and, of course, they rely on intelligent and disciplined obedience from subordinates to achieve their goals. Intelligent obedience requires both mutual understanding and some sympathy, both of which are enhanced by dialogue between leader and follower. The superior who values the perspective of subordinates must create a space in which it is possible for subordinates to express doubt or disagreement without prejudice, and without the superior fearing a loss of authority and the intermediate distance between levels of responsibility that enables objectivity and enhances authority. Frequently, discourse can produce better-informed and more nuanced solutions. At least it can enhance mutual understanding. For the follower, forthright communication is an obligation of loyal subordination and discipline.

Subordinate officers who have the opportunity to address their superiors must be both willing and knowledgeable about how to speak truth to authority. Sometimes this carries risks. Retired Air Force General Charles Boyd told the 2006 graduating class of the Air University that he knew it was hard to oppose "strong willed bosses, even when you're certain you are right. . . . But," he went on, "this is the only professional—indeed ethical—course available to you. In the autumn of your years . . . you will be proudest of those times you took the risk to do the right thing and not the expedient. And you will be most ashamed to recall the times you remained silent when you should have stated your mind."[20]

There are some useful techniques to ease entry into a challenging dialogue. Former Chairman of the Joint Chiefs of Staff General Peter Pace, USMC, learned early on to enter dialogue with superiors by asking questions—by seeking the commander's superior insight so as better to understand the issues. Pace also emphasizes the importance of the superior demonstrating his openness to inquiry, acknowledging a good question, and showing willingness to explain how the issues raised appear from the boss's perspective.[21] This takes time, and ought to be the default approach for the senior officer. Sometimes all that time allows is a simple, "you have your orders." The subordinate should understand and accept that reality.

A key element of subordinate success is maintaining a professional demeanor that accepts as an opening premise that the superior commander is guided by good intentions, has greater experience, far wider responsibilities, as well as many sources of information not available to subordinates. That is, after all, one meaning of the corporateness in a profession—reciprocal respect among practitioners. Subordinates who challenge their superiors should be self-aware, prepared to acknowledge their understanding may be incomplete or misinformed, accept that their motivations may be misunderstood, and offer their judgments not as indictments but as an honest attempt to further the common effort. Subordinates must keep in mind that the measure of any specific mission is its contribution to the total effort, not immediate convenience or cost to their particular unit. Sometimes, when confronted with a problematic tasking, a good approach is to offer a better alternative to achieve the same or more productive result, rather than outright rejection of the superior's immediate vision.

Central Virtues

Four basic virtues are central to the character of Armed Forces officers: discipline, courage, competence, and self-sacrifice (sometimes called selfless service).

Discipline is listed first, because the commission and oath to the Constitution call for it. The officer is admonished to obey the orders of the President and superior officers acting in accordance with the laws of the United States. The oath requires submission to and support of the division of authority and responsibility laid down in the Constitution. In the broader society today, discipline seems to be somewhat out of fashion as a limit upon freedom of action, but it is essential to the reliability of a military force.

In his General Order of January 1, 1776, General George Washington wrote:

> [a]*n Army without Order, Regularity & Discipline, is no better than a Commission'd Mob. . . it is Subordination & Discipline (the Life and Soul of an Army) which next under providence, is to make us formidable to our enemies, honorable in ourselves, and respected in the world; and herein is to be shown the Goodness of the Officer* [emphasis added].[22]

One hundred sixty-seven years later, General Dwight D. Eisenhower, in a letter to his son, then a cadet at West Point, wrote: "We sometimes use the term 'soul of an army.' That soul is nothing but discipline, and discipline is simply the certainty that every man will obey orders promptly, cheerfully and effectively."[23]

Obedience may sometimes allow discretion as to detail, but reliable service as a basis of mutual trust will not dispense with enthusiastic compliance. As Eisenhower also wrote, "The one thing you are going to depend upon is a certain knowledge that every soldier in your unit will do what you tell him, whether you are watching him or not."[24]

Courage, of course, is the obvious virtue for the warrior. The Armed Forces officer requires the courage to dare, the courage to endure, the courage to keep one's head in the midst of chaos and uncertainty, "when everyone around is losing theirs."[25] The officer requires the courage to decide and act. Physical courage is a sine qua non for the officer, as war is a dangerous business. But equally important is moral courage. This is the courage to speak truth to authority, and the courage to act and then to be accountable—the courage to order another Soldier, or a lot of other Soldiers, Marines, Sailors, Airmen, or Coastguardsmen, to take some action that will cost some, sometimes many of them, their lives.

Competence in the necessary skills of whatever position held is the virtue that, with discipline, makes the Armed Forces a reliable instrument providing security to the Nation and leads to successful accomplishment of missions assigned. It reflects the expertise that is the basis of the officer's claim for professional status and the grant of authority for discretionary application of the Armed Forces officer's unique skills. An incompetent force is a threat to the Nation's security. An incapable officer, even one with an otherwise matchless character, is a threat to the Nation and to the force in which he or she serves.

To ensure a competent force, Armed Forces officers have the dual responsibility of training those under their authority so they are prepared on the day of battle, and engaging in continuous personal learning by study and reflection so they themselves are fit to command when that day arrives. In the American Civil War, Union Brigadier General C.F. Smith, an old regular, summed up his philosophy for Lew Wallace, a Union general and later the author of *Ben Hur*, when Wallace asked his advice about a proffered promotion:

> *Battle is the ultimate to which the whole life's labor of an officer should be directed. He may live to the age of retirement without seeing a battle; still, he must always be getting ready for it exactly as if he knew the hour of the day it is to break upon him. And then, whether it come late or early, he must be willing to fight—he must fight.*[26]

To be considered professionals, officers must be expert in their job. They must continually extend their technical and applicatory knowledge and the skills upon which their authority in their organization and value to the Service rest. In the early days of the Army School of Advanced Military Studies, students immersed in the Howard and Paret translation of *On War* and a hundred other books, received two important pieces of advice at the end of their year of study. Brigadier General Huba Wass de Czege, the founding director, told them that the first thing they needed to do when they got to their unit was to qualify as "expert" with their weapon and "max" the PT (physical training) test. General Barry McCaffrey, one of his generation's most distinguished combat officers, told the graduates coming to his command that when he called them to the operations map in his tactical operations center, it was not for a discussion of Clausewitz. He expected the graduates to be experts in their practical business—to be competent in the discretionary application of the profession's specialized knowledge.

Finally, there is *self-sacrifice*, or selfless service. Self-sacrifice is a measure of commitment to a cause as opposed to a simple search for martyrdom. In 1980, Herman Wouk delivered his second Spruance Lecture at the Naval War College, titled "Sadness and Hope: Some Thoughts on Modern Warfare." He offered reflections on Israeli Colonel Jonathan Netanyahu, commander of the Israeli raid on Entebbe, who was the only Israeli fatality in that operation, and Commander Walter Williams, a U.S. Navy officer Wouk had met on an earlier trip to the college. Williams, a pilot, had been killed at sea in a training accident. Wouk reflected on the apparent futility of Williams's death compared to Netanyahu's, pointing out that Williams had survived numerous combat missions over Vietnam. Speaking of Williams, Wouk asked, "What did he achieve with this accidental death in routine operations?" He answered:

> *I'll tell you what he did—he served. He was there. This man of the highest excellence submerged himself, his life, in this big destructive machine which is our solace and our protection, knowing full well that whether he flew combat missions or routine operations he was at risk. He gave up all the high-priced opportunities in this rich country . . . and he served.*[27]

Notes

[1] *The Armed Forces Officer* (Washington, DC: U.S. Government Printing Office, 1950), "Introduction," ii.

[2] Ibid., 164–165.

[3] Sir John Winthrop Hackett, *The Profession of Arms* (New York: Macmillan, 1983), 141.

[4] Julia Annas, *Intelligent Virtue* (London: Oxford University Press, 2011, 2013), 8-9.

[5] Aristotle, *The Ethics of Aristotle: The Nicomachean Ethics*, trans. J.A.K. Thomson (New York: Penguin Books, 1953), 91.

[6] United States Naval Academy, "Honor Concept," available at <www.usna.edu/About/honorconcept.php>.

[7] Marine Corps Warfighting Publication 6-11, *Leading Marines* (Washington, DC: Headquarters Department of the Navy, November 27, 2002), 35, available at <www.marines.mil/Portals/59/Publications/MCWP%206-11%20Leading%20Marine.pdf>. Mark Osiel, quoted in Shannon E. French, *Code of the Warrior: Exploring Warrior Values Past and Present* (Lanham, MD: Rowman & Littlefield, 2003), 14.

[8] Sir John Winthrop Hackett, "The Military in the Service of the State," 1970 Harmon Memorial Lecture at the United States Air Force Academy, in *The Harmon Memorial Lectures in Military History*, 1959-1987, ed. Harry R. Borowski (Washington, DC: Office of Air Force History, 1988), 522.

[9] Lieutenant General H.R. McMaster, "The Warrior Ethos at Risk: H.R. McMaster's Remarkable Veterans Day Speech," November 18, 2014, posted by Janine Davidson, Council on Foreign Relations *Defense in Depth* blog, available at <http://blogs.cfr.org/davidson/2014/11/18/the-warrior-ethos-at-risk-h-r-mcmasters-remarkable-veterans-day-speech/>.

[10] Christopher Coker, *The Warrior Ethos: Military Culture and the War on Terror* (London: Routledge, 2007), 136–137.

[11] George C. Marshall, "The Challenge of Command: An Address to the Graduates of the First Officer Candidate School, Fort Benning, Georgia, September 18, 1941," *Selected Speeches and Statements of General of the Army George C. Marshall*, ed. H.A. DeWeerd (Washington, DC: The Infantry Journal, 1945), 221.

[12] James Davidson Hunter, *The Death of Character: Moral Education in an Age without Good or Evil* (New York: Basic Books, 2000), 16.

[13] George C. Marshall, "The Challenge of Command," 220–221.

[14] Douglas MacArthur, "Thayer Award Acceptance Speech," West Point, NY, May 12, 1962, available at <www.americanrhetoric.com/speeches/PDFFiles/Douglas%20 MacArthur%20-%20Thayer%20Award%20Address.pdf>.

[15] Carl E. Mundy, Jr., Commandant of the Marine Corps, January 3, 1995, foreword, in Marine Corps Warfighting Publication 6-11.

[16] Ibid., 22.

[17] See, for example, the work of the Character Education Partnership at <www. character.org>; or the Character Counts program of the Josephson Institute at <http://charactercounts.org/home/index.html>.

[18] John W. Vessey, Jr., "A Concept of Service," *The Naval War College Review* 51, no. 1 (Winter 1998), 158, available at <www.usnwc.edu/Publications/Naval-War-College-Review/Archivedissues/1990s/1998-Winter.aspx>.

[19] Phillip S. Meilinger wrote an excellent discussion of the Rules of Followership. See Phillip S. Meilinger, "The Ten Rules of Good Followership," article prepared for Air University Publication 2-4, *Concepts for Air Force Leadership*, available at <www. au.af.mil/au/awc/awcgate/au-24/meilinger.pdf>.

[20] General Charles G. Boyd, USAF (Ret.), Air University Graduation, May 25, 2006, "Remarks by General Charles G. Boyd, USAF (Ret.)", available at http://www. au.af.mil/au/aul/img/boyd_speech_auGrad_2006.pdf>.

[21] Peter Pace, "General Peter Pace: The Truth as I Know It," *Ethix*, October 1, 2008, available at <http://ethix.org/2008/10/01/the-truth-as-I-know-it>.

[22] General George Washington, General Orders, January 1, 1776, in George Washington, *Writings* (New York: The Library of America., 1997), 196.

[23] Dwight D. Eisenhower to John S.D. Eisenhower, May 22, 1943, Number 1016 in *The Papers of Dwight David Eisenhower: The War Years*, vol. II, ed. Louis Galambos (Baltimore, MD: Johns Hopkins University Press, 1970), 1151–1152.

[24] Ibid.

[25] From the poem "If," with apologies to Rudyard Kipling.

[26] Bruce Catton, *This Hallowed Ground* (New York: Pocket Books, 1961, 1976), 87–88.

[27] Herman Wouk, "Sadness and Hope: Some Thoughts on Modern Warfare: A Lecture Given on 16 April 1980 at the Naval War College," *The Naval War College Review* 32, no. 5/Sequence 281 (September–October 1980), 11, available at <www. usnwc.edu/Publications/Naval-War-College-Review/Archivedissues/1990s/1998-Winter.aspx>.

CHAPTER FOUR

The Officer at Work: The Ethical Use of Force

Being a person of virtue and good character is integral to being a professional. It is necessary, but not sufficient. A physician may be a person of unassailable character, but to be fully successful in the practice of medicine, she will need to know and be able to apply both the technical skills and the ethical principles that inform and guide such matters as end-of-life treatment options, or whether to be fully truthful with a terminal patient. An attorney might be a person of unquestionable virtue, but he will need to know and be able to apply the principles and rules that spell out the limits on what he is permitted to do in prosecuting a defendant on behalf of the United States, that is, to recognize those actions that might violate his obligations as an officer of the court.

Similarly, in addition to embodying and practicing "soldierly virtues," the military professional, especially the officer, must know and be able to apply the principles and rules that inform and govern the various types of work in which the military engages. The most obvious and important, indeed *defining*, work of the profession of arms is the conduct of war, more broadly the use of deadly force on behalf of the Nation.

Centuries of tradition and law provide that war, in fact any use of force by professional militaries, is a *rule-governed* activity. Those rules have been derived from what Michael Ignatieff called the "warrior's honor" in his book of the same title:

> *While these codes vary from culture to culture, they seem to exist in all cultures, and their common features are among the oldest artifacts of human morality. . . . As ethical systems, they were primarily concerned with establishing the rules of combat and defining the system of moral etiquette by which warriors judged themselves to be worthy of mutual respect.*[1]

The basic notion of the warrior's honor, that not all killing and destruction are legitimate, is nearly universal, transcending historical periods and cultures. It serves more than one purpose: distinguishing between those who fight honorably and those who do not, regulating acceptable weapons and practices, and defining acceptable treatment of prisoners and the wounded. Only men and women who fight under such codes are members of an honorable profession. They are soldiers and warriors, and can proudly call themselves such. Those who fight outside or without such codes are not members of an honorable profession. With no code to inspire and bind them, they are, instead, barbarians, pirates, or criminals. "For war, unconstrained by honor and high moral principle, is quickly reduced to murder, mayhem, and all the basest tendencies of mankind."[2]

The Just War Tradition and the Law of Armed Conflict

For sons and daughters of the Western heritage, the primary ethical code governing the resort to and conduct of war is the Just War tradition. David Fisher, a retired British civil servant, wrote:

> *Just War . . . is not based on a fixed body of doctrine but is rather a tradition that has evolved over the centuries and is still evolving in response to the changing circumstances and nature of war. . . . But within this shifting tradition there is a reasonably settled set of core principles, built up and crafted over the centuries, which are designed to provide guidance to our thinking about war.*[3]

This moral tradition has many kinds of roots—in philosophy, theology, law, the practice of statecraft, and military codes such as chivalry.

The moral principles that inform and govern the member of the profession of arms regarding war and the use of force are not inherently inconsistent with more practical military considerations. While there often is tension between the demands of strategy or tactics and the demands of ethics and law, there is no necessary or fundamental conflict between the two, viewed from the perspective of a successful final outcome.

The Just War ethical principles are customarily divided into two parts: *jus ad bellum,* which informs and governs the decision to go to war or to resort to the use of armed force; and *jus in bello,* which informs and governs the use of force on the battlefield. Michael Walzer has aptly distinguished between the two, saying that the former has an adjectival character (is this a *just* war?) and the latter an adverbial character (is this war being waged *justly*?).[4] The moral burden of *jus ad bellum* falls primarily on political leaders, because they are the ones who make the decision to go to war. The military are not off the hook entirely, however, because they provide military counsel to political leaders, on matters such as feasibility (related to the *jus ad bellum* criterion of probability of success) and on costs and risks (related to the *jus ad bellum* criterion of proportionality) involved in any proposed action.

On the other hand, the moral burden of the *jus in bello* falls primarily on the military. They are the ones who conduct war. Political leaders are not entirely off the hook here either. The means they provide, and their guidance on acceptable actions, can directly or indirectly influence rules of engagement, which will govern the limits imposed on the men and women on the fighting line, and the corresponding risks to Servicemembers these rules entail.

Most formulations of *jus ad bellum* include the following criteria:

- Just cause—the reason for going to war must be sufficiently grave.
- Competent authority—only the duly constituted civil authorities may order the initiation of war.
- Right intention—those initiating war must not have a hidden or ulterior motive.
- Probability of success—there should be a reasonable prospect of success.

- Proportionality—the harm that will be done in the war must not exceed the good that will be accomplished.
- Last resort—war should be undertaken only if nonviolent means to resolve the issue have failed or are unlikely to succeed.

As stated above, consideration of the ethical principles governing the resort to war often parallels consideration of strategy and policy. While ethicists might speak of *probability of success* in assessing whether a proposed war would be just or not, military planners would be deeply engaged in assessing the *feasibility* of various courses of action under consideration.

Similarly, considerations of the *jus ad bellum* criterion of proportionality will be not unlike the military planner's calculations of costs, risks, and the unintended consequences of assigned limitations and the long-term effect of the neglect of such limitations. For example, in 1956, in deciding whether the United States should take military action in support of Hungarians who rose up against Soviet occupation of their country, President Eisenhower probably calculated that the risks of a U.S.-Soviet nuclear war in Europe vastly outweighed any good that might be accomplished by any conceivable U.S. military intervention on behalf of the lightly armed, but valiant citizens of Budapest facing off against Soviet tanks. An ethicist looking at the same decision in terms of *proportionality* in the *jus ad bellum* sense would probably also have concluded that the harm that would ensue from a possible U.S. or Soviet use of nuclear weapons would likely be far greater than the good to be accomplished in helping an oppressed people regain their freedom.

According to *jus in bello*, for a war to be conducted justly it must, *inter alia*, meet two basic criteria:

- discrimination, which deals with intentions[5]
- proportionality, which deals with consequences.

The principle underlying *discrimination* (between those who are legitimate targets of attack—combatants—and those who are not legitimate targets of attack—noncombatants) is noncombatant immunity: noncombatants may never be the object of an intentional direct attack.

This is the realm of *intentions*. The soldier may not *intend to kill or harm* noncombatants. In Just War terms (though the legal terminology may be different), noncombatants include not only civilians caught up in the maelstrom of war, but unresisting enemy soldiers who are wounded and out of the fight, and those who have surrendered and been taken prisoner. In ethical terms, combatants include not only most military personnel, but also civilians actively engaged in the war effort (for example, delivering ammunition to the front lines or taking up weapons themselves). Walzer argues that distinction is based on *rights,* including the right not to be attacked:

> *We try to draw a line between those who have lost their rights because of their warlike activities and those who have not. On the one side are a class of people, loosely called "munitions workers," who make weapons for the army or whose work directly contributes to the business of war. On the other side are all those people who, in the words of the British philosopher G.E.M. Anscombe, "are not fighting and are not engaged in supplying those who are with the means of fighting."*[6]

Proportionality is the realm of the *consequences* of military operations. It says that the harm likely to be done in any particular military operation should not outweigh the good likely to be accomplished by that military operation; that is, it must not be disproportionate to the legitimate gains to be achieved by the military operation. Proportionality acknowledges, in effect, that some noncombatants may be harmed or killed in a military operation, not by direct intent of those conducting it, but as accidental, unintended results, what are often referred to as "collateral damage."

In their 1983 pastoral letter *The Challenge of Peace*, the U.S. Catholic bishops posed the problem this way: "When confronting choices among specific military options, the question asked by proportionality is: once we take into account not only the military advantages that will be achieved by using this means but also all the harms reasonably expected to follow from using it, *can* its use still be justified?"[7] In its discussion of proportionality and discrimination, the U.S. Army and Marine Corps *Counterinsurgency Field Manual* (2007) states:

> *In COIN* [counterinsurgency] *operations, the number of civilian lives lost and property destroyed needs to be measured against how much harm the targeted insurgent could do if allowed to escape. If the target in question is relatively inconsequential, then proportionality requires combatants to forego severe action, or seek noncombative means of engagement.*[8]

Two slightly different vocabularies, but a similar logic, are at work in these very different documents with very different sets of authors.

Implicit in the *jus in bello* principles of discrimination and proportionality is the notion that soldiers, especially their officers, are responsible for noncombatants in their area of operations. Michael Walzer captured this idea in the 1980 essay "Two Kinds of Military Responsibility." In it, he argued that in addition to what he called "the hierarchical responsibilities of the officer," that is, to those above and below in the chain of command, there is another set of responsibilities, which are nonhierarchical. "As a moral agent, [the officer] is also responsible *outward*—to all those people whose lives his activities affect."[9] Those noncombatants, Walzer argues, have rights, including the right not to be harmed or killed, at least not intentionally. This line of thinking is not confined to political philosophers such as Walzer. One can see it in official military publications, such as the *Counterinsurgency Field Manual*. In this document, the argument is made in the context of counterinsurgency, but the point has broader applicability: senior leaders, the manual says, must "assume responsibility for everyone in the [area of operations]."[10]

While the Just War tradition has its primary roots in the West, its underlying principles have been enshrined in the Law of Armed Conflict (LOAC) and International Humanitarian Law (IHL), bodies of international law that are considered binding across the globe for nations with different philosophical, religious, and cultural traditions. Indeed the U.S. Army's *Law of Armed Conflict Deskbook* cites Aristotle, Cicero, Thomas Aquinas, and other philosophers in its discussion of the roots and evolution of LOAC.[11] These principles are formalized in international law, ratified in treaties, and embodied in national military codes. The principles provide a common ground for distinguishing warriors from barbarians, and honorable soldiers from war criminals.

Acts that violate this code offend the human conscience. Thus for the military member, the principles underlying the Just War tradition and the laws of war are not mere abstractions. The importance of these principles to the profession of arms is seen most clearly in the fact that a U.S. Servicemember who violates the LOAC or IHL may be held criminally liable for war crimes and court-martialed under the Uniform Code of Military Justice.

The American military, by and large, depends on material solutions to strategic and tactical problems. Expenditure of material, sometimes massive expenditure, is used to reduce risk to American combatants. Unchecked, however, this practice can lead to actions entirely disproportionate to the intended gains or potential losses both tactically and strategically. Aside from being unnecessarily costly in economic terms, there is in this practice something fundamentally opposed to American values when it leads to unnecessary casualties to noncombatants, or former enemy combatants now under U.S. control. In conflicts that depend upon support, or at least acquiescence, by the local population, failure to discriminate can quickly turn liberators into invaders and impose significant additional manpower demands on local commanders. Portrayal of unnecessary killing in support of nonexistential political goals can produce opposition on the home front that is reflected ultimately in loss of public support for the effort. At the same time, disproportionate friendly losses attributed to overly restrictive rules of engagement have a detrimental effect on soldier morale and also can impact public support. This imposes a requirement on military professionals to discriminate in the use of force, for both practical and ethical reasons.

Though Just War and LOAC limits can impose some immediate risk to soldiers on the firing line, they can, and often do, point toward imposing reciprocal limits on war at the cutting edge, and more broadly on the resort to and the use of military force at the highest levels.

Examples from the American Profession of Arms

The idea that war is a rule-governed activity is deeply embedded in the American psyche and in the DNA of American practitioners of the profession of arms. The idea has been reinforced over the centuries.

In the 18th century, the idea that war is to be rule-governed was made clear even before the Declaration of Independence. As Yale Law professor John Fabian Witt noted:

> *In June 1775, as the War of Independence got underway, the Continental Congress wrote the laws of war into George Washington's commission as commander in chief of the Continental Army. "You are to regulate your conduct in every respect," the Congress told Washington, "by the rules and discipline of war."*[12]

Later, the author and signers of the Declaration of Independence included in their bill of particulars against King George items related to military actions under his purview:

- "That he has been protecting [British soldiers], by a mock Trial, from punishment for any Murders which they should commit on the Inhabitants of these States."
- "He has plundered our seas, ravaged our Coasts, burnt our towns, and destroyed the lives of our people."
- "He is at this time transporting large Armies of foreign Mercenaries to compleat the works of death, desolation, and tyranny, already begun with circumstances of Cruelty & perfidy scarcely paralleled in the most barbarous ages, and totally unworthy of the Head of a civilized nation."[13]

In the 19th century, during our bloodiest war, the Civil War, President Abraham Lincoln ordered the publication of General Orders No. 100, drafted by Francis Lieber, a law professor and student of the laws of war, and revised by a Board of Officers. Among its provisions were:

- "15. Men who take up arms against one another in public war do not cease on this account to be moral beings, responsible to one another, and to God."
- "16. Military necessity does not admit of cruelty, that is, the infliction of suffering for the sake of suffering or for revenge,

nor of maiming or wounding except in fight, nor of torture to extort confessions. It does not admit of the use of poison in any way, nor for the wanton destruction of a district. . . . in general, military necessity does not include any act of hostility which makes the return to peace unnecessarily difficult."

- "22. [A]s civilization has advanced during the last centuries, so has likewise steadily advanced, especially in war on land, the distinction between the private individual belonging to a hostile country and the hostile country itself, with its men in arms. The principle has been more and more acknowledged that the unarmed citizen is to be spared in person, property, and honor as much as the exigencies of war will admit."
- "25. [P]rotection of the inoffensive citizen of the hostile country is the rule; privation and disturbance of private relations are the exceptions."[14]

In the 20th century, General of the Army Douglas MacArthur stated, "The soldier, be he friend or foe, is charged with the protection of the weak and unarmed. It is the very essence and reason of his being . . . [a] sacred trust."[15] Moral soldiers do not harm prisoners, and they accept additional risk to safeguard the helpless. As S.L.A. Marshall wrote in the original edition of *The Armed Forces Officer*, "The barbarian who kills for killing's sake and who scorns the laws of war at any point is repugnant to the instincts of our people, under whatever flag he fights."[16]

In the early 21st century, then-Major General James N. Mattis sent a letter to all those in the First Marine Division (Reinforced) as they were about to cross the line of departure in the Iraq War, telling them in part:

> *Our fight is not with the Iraqi people, nor is it with members of the Iraqi army who choose to surrender. While we will move swiftly and aggressively against those who resist, we will treat all others with decency, demonstrating chivalry and soldierly compassion for people who have endured a lifetime under Saddam's oppression.*[17]

The Challenge for the Officer

The challenge for the division commander is to ensure that these high sentiments have credibility and vitality four or five levels below, in the squads and sections where combat occurs. The challenge and the moral danger for the soldier who fights under such a code are that in the heat and fury of combat, Clausewitz's fog and friction, there are powerful forces, the "forces of moral gravity," which tend to drag the soldier down to the enemy's level. Not least among these is a well-developed survival instinct. At the sharp end, restrictions on the range of acceptable actions often carry increased personal risk to the warfighters that must be weighed against other considerations that may be governing their actions.

The enemy's reciprocity of respect for humanitarian codes—or lack of reciprocity—will weigh heavily with the members of the infantry squads. Notwithstanding the enemy's conduct, the moral and legal codes that should govern the conduct of American military professionals are those they bring with them to the war, not those the enemy brings to the fight. They must resist being dragged down to the level of an unscrupulous enemy, no matter how strong the temptation. To do this, they need help. Resisting the forces of moral gravity is the work of ethics, the law, training, education, leadership at all levels, and command. Fundamentally, it is a matter of discipline.

Attention to the laws of war is the special responsibility of officers. Junior officers, dedicated to ensuring the lowest possible losses to their troops, must be reminded that it is neither their responsibility, nor within their abilities, to make combat operations risk free, especially by compromising standards of legal conduct. Achieving the proper balance between mission accomplishment and risk to noncombatants and soldiers is one of the reasons why Presidents place "special trust and confidence" in the Armed Forces officer. The more forces that pull against the forces of moral gravity, the less the likelihood that individual soldiers will succumb to that downward pull. These stabilizing influences, which must be implanted before the battle, are the responsibility of commanders and other leaders. They require continuous and deliberate inspection and tending.

It is particularly the officer's duty to see that Servicemembers are not compromised by unworthy actions, even in the heat of battle. The

U.S. Army and Marine Corps *Counterinsurgency Field Manual* places a demanding ethical burden on leaders, a burden that falls most heavily on officers, especially commanding officers. These officers must

- "provide the moral compass for their subordinates."[18]
- "work proactively to establish and maintain the proper ethical climate of their organizations."[19]
- "serve as a moral compass."[20]
- "maintain the 'moral high ground' in all their units' deeds and words."[21]
- "not allow subordinates to fall victim to the enormous pressures associated with prolonged combat."[22]
- "establish an ethical tone and climate that guards against the moral complacency and frustrations that build up in protracted . . . operations."[23]

Understanding all these principles, and being able to apply them in practice, are demanding tasks, requiring both classroom learning and frequent practical field training that confronts leaders with the dilemmas of restrictive rules of engagement. Professional military education, especially officer education, plays a central role in this process. What is taught and learned will vary across the career spectrum of military schools and individual experience in training or operations.

For those in officer accession programs (Service academies, Reserve Officers Training Corps, and officer candidate schools), as well as for junior officers, it is critical to master, and indeed internalize, the *jus in bello* aspects of the ethics of the use of force. It is junior officers leading and commanding at the tactical level who will be expected to make the critical decisions, often involving risk of injury and death, without much opportunity for on-the-spot reflection, and usually without the benefit of significant combat experience. They will be responsible and accountable as well for the decisions made by those they command, often men and women of greater age and more experience, who may be more emotionally engaged in the immediate problem. If the first time an officer thinks about the ethical aspects of the use of force is in combat, under fire, the outcomes for the officer, the troops, and innocent noncombatants in the area are likely to be more unfortunate than they might otherwise be.

As officers rise in rank, and assume positions of greater responsibility both to advise civil superiors and guide subordinate conduct, acquiring a working understanding of *jus ad bellum* is also valuable, and is sometimes necessary, depending on the jobs they hold. Those serving on the Joint Staff, in the Office of the Secretary of Defense, in the State Department, on the staffs of Combatant Commands, or on the National Security Council staff, may well be engaging directly, or indirectly, with senior civilian officials charged with formulating recommendations, or even making decisions, regarding the use of force. At that level, in particular, having a working knowledge of the vocabularies and logics of ethics, and of strategy and policy, can facilitate making recommendations or decisions that are both ethically sound and strategically wise.

To reiterate what was said at the beginning of this chapter, being of good character and embodying the right military virtues are essential for success in the most important work the military does. However, not having an intimate understanding of the relevant ethical principles, or lacking the practiced ability to apply them in the real world, may leave an officer less than ideally prepared, or maybe not even adequately prepared, to successfully navigate the challenges of those life-and-death responsibilities thrust upon him or her. Thus it is incumbent upon the Armed Forces officer to master these principles and legal provisions, to apply them in practice, and to instill them in subordinates at all levels.

Notes

[1] Michael Ignatieff, *The Warrior's Honor* (New York: Henry Holt, 1997), 116-117.

[2] Colonel John R. Allen, USMC, "Commandant's Intent," February 19, 2002 version (unpublished), 3.

[3] David Fisher, *Morality and War* (Oxford: Oxford University Press, 2011), 64.

[4] Michael Walzer, *Just and Unjust Wars*, 2nd ed. (New York: Basic Books, 1992), 21.

[5] The term *discrimination* is traditionally used in Just War discourse. The counterpart term in the legal realm is *distinction*.

[6] Walzer, *Just and Unjust Wars*, 145. In his endnote on Anscombe, Walzer cites the following: "G.E.M. Anscombe, *Mr. Truman's Degree* (privately printed, 1938), p. 75; see also 'War and Murder' in *Nuclear Weapons and Christian Conscience*, ed. Walter Stein (London, 1963)."

[7] *The Challenge of Peace: God's Promise and Our Response*, Pastoral Letter on War and Peace by the National Conference of Catholic Bishops (Washington, DC: United States Catholic Conference, May 3, 1983), 46.

[8] U.S. Army Field Manual No. 3-24/Marine Corps Warfighting Publication No. 3-33.5, *Counterinsurgency Field Manual* (Chicago: University of Chicago Press, 2007), 247–248.

[9] Michael Walzer, *Arguing About War* (New Haven: Yale University Press, 2004), 24–25.

[10] *Counterinsurgency Field Manual*, 239.

[11] *Law of Armed Conflict Deskbook* (Charlottesville, VA: International and Operational Law Department of the Judge Advocate General's Legal Center and School, 2014), 8–12.

[12] John Fabian Witt, *Lincoln's Code: The Laws of War in American History* (New York: Free Press, 2012), 15.

[13] *The Declaration of Independence and the Constitution of the United States of America* (Washington, DC: National Defense University Press, 1994), 33–35.

[14] Witt, 377–378.

[15] As quoted in Walzer, *Just and Unjust Wars*, 317.

[16] *The Armed Forces Officer* (Washington, DC: U.S. Government Printing Office, 1950), 15.

[17] Commanding General Message to All Hands, March 2003, available at <www.thehighroad.org/archive/index.php/t-14554.html>, April 11, 2015.

[18] *Counterinsurgency Field Manual*, 237.

[19] Ibid., 238.

[20] Ibid., 239.

[21] Ibid.

[22] Ibid., 240.

[23] Ibid.

CHAPTER FIVE

The Officer at Work: Leadership

. . . before it is an honor, leadership is trust;
Before it is a call to glory,
Leadership is a call to service;
. . . before all else, forever and always, leadership is a willingness to serve.

—Father Edson Wood, OSA, Cadet Catholic Chaplain
Invocation at Assumption of Command by BG Curtis Scaparrotti, Commandant of Cadets, U.S. Military Academy
August 11, 2004

Leadership—convincing others to collaborate effectively in a common endeavor—is the primary function of *all* Armed Forces officers. Only a few officers are commanders at any particular moment, but every officer is a leader. Indeed the Army and Marine Corps insist that leadership is the common responsibility of every Soldier and Marine.[1] The Air Force says "Any Airman can be a leader and can positively influence those around him or her to accomplish the mission."[2] A consequence is that almost every officer considers himself or herself good at leadership, but perspectives on method differ depending on individual circumstances and experiences. This chapter discusses leadership from four different but overlapping viewpoints: accomplishing the mission and taking care of the troops; three concepts of leadership; Service approaches; and "tribal wisdom," views of leadership expressed by senior professionals.

Accomplishing the Mission and Taking Care of the Troops

Leaders are expected to guide their followers to mission success at least possible cost. Lord Moran, who served as a medical officer on the Western Front in World War I, and was Churchill's doctor and confidant in World War II, defined leadership as "the capacity to frame plans which will succeed and the faculty of persuading others to carry them out in the face of death."[3] Moran was skeptical of a requirement for fine character, the honorable virtues, in a leader, but found that a reputation for achieving success was the essential middle term between the ability to formulate a course of action and persuading others to implement it. He believed "phlegm—a supreme imperturbability in the face of death . . . [was] the ultimate gift in war."[4]

In the U.S. Armed Forces, the admonition "Take care of your people" is coupled with the requirement for mission success: "Mission first! People always!" This obligation to care for your people is so ingrained that it serves as an ethical principle for those who lead. Indeed, care of subordinates is called for explicitly by three identical passages of Title 10 U.S. Code: Sections 3583 (Army), 5947 (Marine Corps and Navy), and 85831 (Air Force). The statutory "Requirement for Exemplary Conduct" mandates, among other things, that "all commanding officers and others in authority . . . be vigilant in inspecting the conduct of all persons . . . under their command"; that they "guard against and suppress all dissolute and immoral practices"; and that they "take all necessary and proper measures . . . to promote and safeguard the morale, the physical well-being, and the general welfare of the officers and enlisted persons under their command or charge."[5] Still, individual competence remains the first desideratum of Armed Forces officers. As officer-scholar Harold Winton has written: "In war, raw professional competence is a much better harbinger of concern for one's subordinates than is either humility or approachability."[6]

Taking care of the troops means attending to their personal needs—physical, mental, and spiritual—and, to a great extent, to their families' needs as well. It also means treating everyone with dignity and respect. American Soldiers, Marines, Sailors, Airmen, and Coastguardsmen are not hirelings, but professionals. Leaders treat people—subordinates, peers, and superiors alike—with dignity and respect. This is both an institutional norm in every Service and another guiding ethical principle for Armed Forces officers.

Eugene B. Sledge, a teenage Marine mortarman in some of the heaviest fighting in the Pacific during World War II, remembered his company commander, Captain Andrew A. "Ack Ack" Haldane, this way:

> *Captain Haldane was the finest and most popular officer I ever knew. . . . Although he insisted on strict discipline, the captain was a quiet man who gave orders without shouting. He had a rare combination of intelligence, courage, self-confidence, and compassion that commanded our respect and admiration. . . . While some officers . . . thought it necessary to strut or order us around to impress us with their status, Haldane quietly told us what to do. We loved him for it and did the best job we knew how.*[7]

Taking care of the troops also means training and educating subordinates for the demands and challenges of their individual jobs and unit missions. In its fullest sense, individual development means going beyond the immediate requirements of the job and the mission, to helping subordinates grow in their own careers, preparing them for higher rank, for greater responsibility, and most especially for current and future leadership of their own troops. A good leader leads, and a great leader develops other leaders. In 1921, the legendary Commandant of the Marine Corps, Major General John A. Lejeune, put his own distinctive stamp on the quality of leadership he expected of Marine officers:

> *The relation between officers and enlisted men should in no sense be that of superior and inferior nor that of master and servant, but rather that of teacher and scholar. In fact, it should partake of the nature of the relationship between father and son, to the extent that officers, especially commanding officers, are responsible for the physical, mental, and moral welfare, as well as the discipline and military training of the young men under their command who are serving the nation in the Marine Corps.*[8]

Three Concepts of Leadership

Leadership may be examined phenomenologically from a number of overlapping perspectives; three currently seem to have particular resonance with military communities:

- Leadership is a human relationship.
- Leadership is a complex of attributes or characteristics that mark successful leaders.
- Leadership is a process.

Leadership is a human relationship between leaders and followers. In contrast to command, which depends on a grant of legal authority, assigned responsibilities, and formal accountability, leadership involves a human bond, a decision by one person to take charge, and corresponding decisions to follow and collaborate by others—followers who submerge their own actions in the vision of the leader. Following may be voluntary, coerced, or negotiated. It may occur simply because one member of the group appears to know what is required right now, when others are confused or hesitant. Followers are the essential complement to the leadership equation.

In an often overlooked 1958 classic about infantry squads, then Colonel William E. DePuy framed a telling epigram, "You can't see an infantry squad—it is an idea that exists only when jointly held by its members."[9] The same could be said about any group acting in harmony to achieve a common end. Instilling, or maintaining, the idea of the group, and following through with collaborative action, are the business of the leader. Another way to put it is this: troops obey because they must; they follow because they want to. They obey superiors; they follow leaders. The obvious is worth stating: an officer must be capable of being both a superior and a leader.

Leadership is a complex of attributes or characteristics that mark successful leaders, men and women who motivate and direct the efforts of others in collaborative enterprises. The premise here is that one simply *is* a leader and the route to development lies in imitation of other successful leaders. This is the more traditional perspective, and it is the one the Marine Corps has maintained most faithfully.

In the foreword to his most developed (1960) version of the of *The Armed Forces Officer*, S.L.A. Marshall characterized the book as dealing with "the two major roles of the officer—as a leader of men, and as a loyal, efficient member of the Nation's defense team."[10] Marshall's books are largely guides to commissioned military leadership, examining what officers do from multiple points of view. While his writing offers numerous useful observations, it would be difficult to distill from it a systematic theory of leadership in the sense of mobilizing individual efforts to achieve shared goals. To the contrary, Marshall's efforts lead largely to a listing of desirable or necessary character traits or attributes of military leaders. These are useful, particularly for novice officers, to help them frame their own place in the profession, and for seasoned leaders to reframe where they stand in their progressive growth.

Marshall's core list of leadership attributes from 1950 onward was:

> *Quiet resolution*
> *The hardihood to take risks*
> *The will to take full responsibility for decision*
> *The readiness to share its rewards with subordinates*
> *An equal readiness to take the blame, when things go adversely*
> *The nerve to survive storm and disappointment and to face toward each new day with the scoresheet wiped clean, neither dwelling on one's successes nor accepting discouragement from one's failures.*[11]

Useful as the attributes approach is, it is inherently not definitive because the various lists often differ in content, sometimes leading to a "battle of the lists." Each advocate thinks his or her list is best, which is to be expected and must be seen in that context.

Leadership is a process, a creative combination of purposeful and identifiable characteristics and behaviors intended to influence others; features and actions that are subject to observation, assessment, evaluation, and correction. This is the view taken by the leadership community in the United States Army. It is discussed in greater depth below.

In fact, these three perspectives are overlapping. Features of one are often accompanied by those of another. Practice of leadership (that is, leadership style) is highly personal and idiosyncratic. It depends on individual disposition, personality, and the leader's understanding of the immediate circumstances. Suitability of a particular style is circumstantial, depending on immediate conditions requiring a collective response, *and* the immediate disposition of potential followers. Followers respond differently, depending on their understanding of the circumstances and their expectations of the leader at any particular moment. Generational differences, which are often significant in defining both leader and follower expectations, must be taken into account.[12] Sometimes troops can be given directions and led by inspiration. Other times, when they are tired or discouraged, they must be driven.[13] Not all leaders are capable by disposition of employing all styles of leadership. Sometimes, senior officers have to pick the right leader at the proper moment for a specific task.

Service Approaches

The Department of Defense does not define *leadership* in Joint Publication 1-02, *Dictionary of Military and Associated Terms*. Moreover, the structure of the leadership experience of Armed Forces officers varies among the Services. For example, Air Force flying officers come to direct leadership of significant numbers of people much later than Army and Marine infantry officers. Some submarine commanders lead fewer troops than an infantry company commander, albeit with a good deal more authority. These structural differences, and the significantly different environments in which the Services operate, undoubtedly influence Service perspectives on the nature and practice of leadership.

Still, Armed Forces officers learn about leadership in a number of common ways. First of all, they have their own experiences organizing and directing others to achieve assigned goals. They learn by doing. Then, they observe others, peers and superiors particularly, and adapt their own practice to take advantage of what they see other successful leaders do, avoiding what they see unsuccessful leaders do. They read about leadership in Service schools, and on their own. They expand their empirical base by reflecting on the experiences of others, often historical leaders like Generals George Washington, U.S. Grant,

William T. Sherman, George Marshall, Dwight Eisenhower, Henry "Hap" Arnold; Admirals Chester Nimitz, Raymond Spruance, and Bill Halsey; and the immortal Marine, General "Chesty" Puller. Fictional accounts like Michael Shaara's *The Killer Angels*, Anton Myrer's *Once an Eagle*, and Herman Wouk's *Caine Mutiny*, influence their thinking, as do the many film representations of air, sea, and land combat, good and bad. Mid-career officers often branch out to sample the mountain of leadership books from various business schools and behavioral science departments that can often be found in airport and other bookstores.

Of all the Armed Forces, the Army seems most devoted to written leadership doctrine. In part, this can be attributed to the fact the Army is a large organization, divided into full-time and significant part-time components. More than the other Services, the Army has to accommodate itself to major periodic expansions in time of crisis and reductions thereafter. Army doctrine, then, takes on a highly structured and positivist form, suitable for formal instruction and institutional application.

For 28 years following World War II, the Army defined leadership as an art and followed a traditional pattern of presenting observed attributes from historical exemplars.[14] Today it treats leadership as a process. The primary Army leadership publication today is Army Doctrine Publication (ADP) 6-22, *Army Leadership*. Its expressed purpose is establishment of "Army leadership principles that apply to officers, non-commissioned officers and enlisted Soldiers as well as Army civilians."[15] The ADP 6-22 defines *leadership* as "*the process* of influencing people by providing purpose, direction, and motivation to accomplish the mission and improve the organization [emphasis added]."[16] The publication defines the Army leader as "anyone who by virtue of assumed rank or assigned responsibility inspires and influences people to accomplish organizational goals."[17] In his foreword to ADP 6-22, General Raymond Odierno, then Army Chief of Staff, wrote, "Being a leader is not about giving orders, it's about earning respect, leading by example, creating a positive climate, maximizing resources, inspiring others, and building teams to promote excellence."[18] In short, Army doctrine is about what leaders must do.

The change in definition from art to process evolved over time. The 1983 Field Manual (FM) 3-22, *Military Leadership*, set the Army on a 20-year path of defining leadership in terms of what a leader had to "Be, Know, and Do" (attributes, knowledge, and action). The course-setting volume began with a study of Colonel Joshua Chamberlain at Gettysburg's Little Round Top as an exemplary model of combat leadership. Current core leadership publications have largely dispensed with historical examples.[19] The absence of exemplars is indicative of an institutional commitment to dependence on the behavioral sciences in formulating leadership doctrine as a tool for helping the Army develop leaders. It seeks to do this by defining leadership in abstract terms on the basis of which observable practices can be taught systematically, observed, evaluated, and then critiqued. Current Army leadership doctrine offers a model that combines abstract attributes of character, presence, and intellect, with observable conduct of leading, developing, and achieving. It acknowledges the importance of followership to leadership. It categorizes leadership by level, as direct, organizational, and strategic; and according to whether it is formal, informal, collective, or situational.[20]

In the Marine Corps, leadership doctrine has been more stable. It remains part of a holistic program of general institutional indoctrination for becoming a Marine. The Marine Corps follows a more traditional pattern of instruction-through-emulation that dates back to the ancients, to Homer and more particularly to Plutarch. Characteristically, Marine Corps Warfighting Publication 6-11 [formerly FMFM 1-0] is given the active title of *Leading Marines* rather than the more impersonal and abstract *Marine Leadership*.

The Marine manual is more inspirational than categorical or dogmatic. It refers to leadership as "the combination of the intangible elements of our ethos and the more tangible elements of our leadership philosophy."[21] Examples for emulation are common. The Marine Corps continues to subscribe to the view of General Lejeune that "leadership is a heritage which has passed from Marine to Marine since the foundation of the Corps . . . mainly acquired by observation, experience, and emulation. Working with other Marines is the Marine leader's school."[22] The core reference remains paragraph 1100 of the *Marine Corps Manual*, which lists three Marine Corps leadership qualities:

inspiration, technical proficiency, and moral responsibility. It quotes General Lejeune's 1921 instruction:

> *Leadership.—Finally, it must be kept in mind that the American soldier responds quickly and readily to the exhibition of qualities of leadership on the part of his officers. Some of these qualities are industry, energy, initiative, determination, enthusiasm, firmness, kindness, justness, self-control, unselfishness, honor and courage. Every officer should endeavor by all means in his power to make himself the possessor of these qualities and thereby to fit himself to be a real leader of men.*[23]

In reflecting on Lejeune's use of "soldier," it is worth remembering that he commanded both Marine and Army units as the commander of the 2nd Division in the American Expeditionary Forces in World War I.

The Air Force and Navy appear more concerned with individual development programs in which leadership techniques are acquired through progressive and varied experiences, including terms of professional education, than with didactic doctrine.[24] The professional organization of the sea Services, the U.S. Naval Institute, publishes a family of officer guides written by notable Navy officers such as Admiral James Stavridis. In May 2014, the Navy established a Naval Leadership and Ethics Center as a command under the Naval War College "to serve as the Navy and [Naval War College's] instrument to provide curriculum development along with assessment to instill fundamental tenets of ethical leadership throughout the Navy."[25] The Air Force leadership manual, published by the Curtis E. LeMay Center for Doctrine Development and Education at Maxwell Air Force Base, offers a framework for thinking about differences in leadership practice at various levels of responsibility, but is not overly concerned with techniques or specific behaviors.[26] Like the Army manual, it also includes reference to followership as a critical element in the leadership "system." The Air Force also has a formal program for mentorship as part of its development program that may be as important as its formal leadership instruction.[27]

"Tribal Wisdom"

In addition to formal leadership doctrine and examples from experience, biographies, and fiction, the Armed Forces also possess a kind of "tribal wisdom" that is passed from generation to generation in formal presentations, shared observation, and experience. The cumulative notions, retained in the institution, seem remarkably similar throughout the several Service tribes. Valuable elements of tribal wisdom are found in public presentations by senior officers and noncommissioned officers.

In 1999, then Rear Admiral Mike Mullen, Director of Surface Warfare, told a class of surface warfare officers that there were certain core attributes required to succeed as a leader in the Navy: "Truthfulness in everything you do; Trustworthiness to follow direction; Demonstration of a capacity for active listening; and Always do your personal best." To these he added what he called "the fundamental goals of a good liberal education: courage, judgment, curiosity and imagination."[28] From all these he synthesized a set of what he called life-skills: integrity, initiative, responsibility (to Sailors, family, and self), establishment of goals, and flexibility. Mullen concluded:

> *The greatest advice I can give you is the oldest of them all in our community: Get out there and walk around. Talk to your Sailors, other junior officers, the chiefs and even the commanding officer because leadership is about "being there." . . . Being there to influence events on the deck plates. . . . Being there to lead—leading your Sailors. . . . Being there for the good times as well as the bad, just as our Navy is there to carry out its mission.*[29]

Nine years later, Admiral Mullen, by then Chairman of the Joint Chiefs of Staff, advised graduating Midshipmen of the Naval Academy to do three things: to learn from their mistakes; to not be afraid to question their seniors—to stand up for what's right; and to accept accountability. "If you are wrong, admit it. If you have erred, correct it. Hold yourselves accountable for your actions." He continued:

> *The quality of our work and our personal conduct say more about who we are and what we stand for than anything else. You should strive to conduct yourself always in such a manner that it can never be said that you demand less of yourself ... or of the men and women in your charge ... than that which is expected of you by your families or your countrymen.*[30]

Much of this tribal wisdom can be summed up in the following five propositions about leadership.

Leadership is a bond of trust. As the epigraph at the head of the chapter reflects, "before it is an honor, leadership is trust." Followers trust leaders to direct their efforts to success at the least necessary cost. Leaders trust followers to comply with their direction. General Sir John Hackett addressed the link between the leader and the led thus: "The leader," he wrote, "has something which the others want and which only he can provide. . . . This something is partly the ability to find an answer to a problem which the others cannot solve. But there is also the power, when difficulties have to be overcome, to help people over them. . . . What the leader has to give is the direction of a joint effort which will bring success."[31]

In a speech to the West Point Class of 2013, General Martin E. Dempsey, then Chairman of the Joint Chiefs of Staff, showed the cadets a photo of an infantry squad in Afghanistan:

> *You've all heard that warfare is changing, technology is taking over, the Army is a thing of the past. But you know, the most sophisticated piece of warfighting equipment in this picture is this squad leader and he hasn't changed all that much really since the days of the Roman legions. Politics are going to change, technologies will change, the enemy will change, but that squad leader won't. And you his leader can't. . . . He is operating because he trusts that that man or woman to his right flank, that rifleman, is protecting him while he does his job. And similarly, that rifleman who is oriented outward is confident and trusts that the squad leader has his back. It doesn't get any more fundamental than trust. And trust is built on confidence in each other. And confidence comes from recognizing the competence,*

> *the character, the quality of each of us . . . trust is the very foundation of our profession. And if you're not living up to earning your part of that equation, you're not living up to being a member of the profession.*[32]

Trust is omni-directional, a mutual vertical relationship between leaders and followers, and a horizontal reciprocal trust among soldiers that those on their left and right will do their part. S.L.A. Marshall quotes Army General James G. Harbord, General Pershing's Chief of Staff, subsequently commander of the 4th Marine Brigade, and briefly commander of the Second Division in World War I. Harbord wrote: "Discipline and morale influence the inarticulate vote that is constantly taken by masses of men when the order comes to move forward—a variant of the crowd psychology that inclines it to follow a leader. But the Army does not move forward until the motion has carried. 'Unanimous consent' only follows cooperation between the individual men in the ranks."[33]

This bond of trust between leader and led is no less important between higher commanders and soldiers on the fighting line. "Don't worry, General. We trust you," a 3rd Armored Division Soldier told Lieutenant General Fred Franks, VII Corps Commander, on the eve of the ground attack in Operation *Desert Storm*.[34] As General Eisenhower told his son John, the leader must be able to count on the organization doing what he directs (see chapter 3, section titled "Central Virtues," for a discussion of *Discipline*).

Senior leaders trust intermediate leaders to translate their orders into meaningful instructions, which they pass on to their subordinates as their own. Soldiers count on the commander's technical competence, on doing his or her best to buffer the troops from the storms above, and ensuring their success at what the commander orders. S.L.A. Marshall has a telling observation about the importance of junior officers as leaders: "even when things are going wrong at every other level, men will remain loyal and dutiful if they see in the one junior officer who is nearest them the embodiment of the ideals which they believe should apply throughout the service."[35]

A leader builds and nurtures trust in an organization both by *being trustworthy* and by *being trusting*. Troops must be able to take the leader's word at face value and have full confidence in his or her technical competence and moral character. The second element of trust is equally essential: troops must know that their leaders have confidence in them and take their word at face value as well. The officer who continually second-guesses the troops, or micro-manages them, will not be leading an organization distinguished by trust, and thus that officer will fail in a primary obligation. As Admiral William Crowe put it when he was Chairman of the Joint Chiefs of Staff, "You cannot run a unit just by giving orders and having the Uniform Code of Military Justice behind you."[36] Coaching, mentoring, and trusting are critical activities of the successful leader.

Leaders set and enforce standards. Military leaders are responsible for getting the most out of their subordinates, and for protecting them from unnecessary burdens, but leaders are not shop stewards. Orders are *their* orders, and standards are *their* standards. They insist on their achievement. John Baynes, a retired British officer and historian, has written:

> *A strictly imposed discipline is not condescending. . . . To allow a soldier to disobey orders is really to insult him. A good man, in any walk of life, knows what he can do, and what he should do. If he fails, he expects the just reward of failure. . . . A man in authority who lets his subordinates get away with poor performance implies in doing so that they and their actions are of no consequence. . . . Tolerance is not only disliked by the soldier for its implication that his efforts do not matter much, but also because it is to some extent an abnegation of duty by his superior.*[37]

S.L.A. Marshall wrote that "the level of discipline is in large part what the officers in any unit choose to make it. . . . *To state what is required is only the beginning; to require what has been stated is the positive end* [emphasis in original]."[38] Leaders never walk past slackness without acting to correct it. They accept responsibility for maintaining high standards and reinforce their regular attainment.

A key requirement of leadership is the obligation to create and sustain a behavioral space that encourages ethical conduct from

Servicemembers acting under or within the leader's authority. This gets back to the warning from General Mattis to Naval Academy midshipmen quoted in chapter 1 that, "you must make certain that your troops know where you are coming from and what you stand for and, more importantly, what you will not tolerate."[39] Setting boundaries of acceptability can be formal, presented as command direction, or it can be as simple as reminding members of the principles of ethical conduct and correcting subordinates for acts of laxness such as using racial epithets to refer to host nation civilians, or using false bravado to encourage aggressiveness. Soldiers learn what is acceptable very much by watching how their superiors react or don't react to what is going on around them.

A major part of setting formal boundaries is their public enforcement. Equally important are the informal methods leaders employ to avoid violations by setting a desirable tone, especially by being aware and alert. Leaders must always be attentive to what is actually taking place in their unit by "being out there," listening actively to junior Servicemembers, both in what they say directly when questioned and what the leader hears them say when they are talking among themselves. Sometimes comments made in humor by one member to another can reveal an ethical laxness that can grow if not corrected. Leaders must attend continually to the ethical space, or it risks being taken over by others with different standards and values. Setting proper boundaries and encouraging ethical behavior protect subordinates from the dehumanizing effects of the combat environment.

Leaders set the example. General Colin Powell said of the relationship between Soldiers and platoon leaders:

> *They will look to you for inspiration, for a sense of purpose. They want to follow you, not be your buddy or your equal. You are their leader. They want someone in charge who they can trust—trust with their lives. They want someone they respect, someone they can be proud of. They want to be able to brag about their lieutenant.*[40]

Officers set the example every day by demonstrating their technical knowledge, their physical conditioning, and their professional appearance and deportment, and particularly by exhibiting a positive attitude in the face of adversity. In conditions of stress, they must maintain a calm demeanor and demonstrate self-possession if they expect the same from their troops. Soldiers will key off of the leader in times of stress. Commander Thomas Buell related a story about Admiral Spruance, whose flagship was hit by a kamikaze off Okinawa in World War II. The staff was unable to find the Admiral and searched for him around the ship. They found him manning a hose in a burning area of the ship with members of a fire control party, applying "leadership on the deck plates," as Admiral Mullen put it.[41] More recently, when the Pentagon was hit by the terrorist attack on 9/11, members of the Army staff finding their way through the dark out of the chaos and carnage remembered looking up and seeing General Jack Keane, a big man and then the Army Vice Chief of Staff and a four-star general, walking calmly into the dark corridor to see where he could help.[42]

Leaders are models of courage, physical and moral. Physical courage is an obvious requirement for military leaders. The leader who is seen to hesitate or lack confidence in battle loses credibility with those who depend on him. Moral courage—the courage to act under conditions of stress, to do what circumstances require and accept responsibility, to give an order and make it stick—is something less commonly addressed. There is, perhaps, no requirement for moral courage greater than sending soldiers into battle. This is true for senior commanders who send forces into harm's way, knowing all will not return, and particularly for junior officers and enlisted leaders who live with and know the people they lead and command personally, as individuals. General Peter Pace, at the time Vice Chairman of the Joint Chiefs of Staff, warned West Point cadets:

> *Your soldiers want to follow you. They want you to be good. They will cling to leaders who care about them. The worst thing that you can do in combat is get yourself killed. It's also the easiest thing to do in combat. . . . As a leader you will have to decide who does what in life and death situations. And I will tell you that you will want to do it yourself. You'll want to do it yourself because A, you know that you know how to do it; and B, it's easier to do it yourself than to send one of your soldiers out and watch them get killed doing what you told them to do.*

> *But you've got more than one soldier, and all of your soldiers are looking to you for leadership. They will do whatever you tell them to do. They do not want you to do it for them. . . . They understand the risks. But if you go do it and you get killed, you have taken away their leadership. And in thinking that you were being self-sacrificing you have really done damage to your unit.*[43]

Notably, General Pace remembered the name of each Marine he lost as a platoon leader in the Battle of Hue in Vietnam; remembering the names of lost comrades is not an uncommon trait among combat leaders. When the decision was made not to nominate Pace for a second term as Chairman of the Joint Chiefs of Staff because of congressional opposition to the policy of the administration he served, many suggested the general should resign before his term was over. He told an audience at the Joint Forces Staff College on June 15, 2007: "I said I could not do that for one very fundamental reason," which was that no Soldier or Marine in Iraq should "think—ever—that his Chairman, whoever that person is, could have stayed in the battle and voluntarily walked off the battlefield."[44] Although it did not occur on the battlefield, Pace's stand was modeling moral courage too, to say nothing of a professional's sense of duty and a personal sense of proportion.

Leaders build and sustain morale. Morale is the combination of pride and collective self-esteem that binds units into organizations greater than the sum of their parts—*esprit de corps,* which S.L.A. Marshall calls "what the unit gives the man in terms of spiritual force translated into constructive good."[45] "*Esprit,*" he writes, "is the product of a thriving mutual confidence between the leader and the led, founded on the faith that together they possess a superior quality and capability."[46] *Esprit* reflects a collective morale, which has its foundation in the individual. Individual morale nurtures the shared determination to prevail, come what may. In his memoir of service with the Indian Army in World War II, novelist and former officer of the Indian Army John Masters quotes a speech on morale given early in the war by Field Marshall "Bill" Slim:

> *In the end every important battle develops to a point where there is no real control by senior commanders. Each soldier feels*

himself to be alone. Discipline may have got him to the place where he is, and discipline may hold him there—for a time. Co-operation with other men in the same situation can help him to move forward. Self-preservation will make him defend himself to the death, if there is no other way. But what makes him go on, alone, determined to break the will of the enemy opposite him, is morale. Pride in himself as an independent thinking man, who knows why he's there, and what he's doing. Absolute confidence that the best has been done for him, and that his fate is now in his own hands. The dominant feeling of the battlefield is loneliness, gentlemen, and morale, only morale, individual morale as a foundation under training and discipline will bring victory.[47]

S.L.A. Marshall, like Slim, was a student of morale. Before he wrote *The Armed Forces Officer*, he wrote *Men Against Fire: The Problem of Battle Command*, which many still consider a classic study of leadership in combat.[48] In his initial edition of *The Armed Forces Officer*, Marshall paired morale with discipline and argued that the second derived from the first. "The Moral strength of an organic unity," Marshall wrote, "comes from the faith in [the] ranks that they are being wisely directed and from faith up top that orders will be obeyed." Discipline he defined as "simply that course of conduct which is most likely to lead to the efficient performance of an assigned responsibility."[49] To achieve moral strength requires effective leadership:

The art of leadership, the art of command, whether the forces be large or small, is the art of dealing with humanity. Only the officer who dedicates his thought and energy to his men can convert into coherent military force their desire to be of service to their country. . . . Diligence in the care of men, administration of all organizational affairs according to a standard of resolute justice, military bearing in one's self, and finally, an understanding of the simple facts that men in a fighting establishment wish to think of themselves in that light and that all military information is nourishing to their spirits and their lives, are the four fundamentals by which the commander builds an all sufficing morale in those within his charge.[50]

Notes

[1] Marine Corps Warfighting Publication (MCWP) 6-11, *Leading Marines* [formerly FMFM1-0] (Washington, DC: Headquarters Department of the Navy, November 27, 2002), 33; and Les Brownlee and Peter Schoomaker, "Serving a Nation at War," *Parameters* (Summer 2004), 13.

[2] *Volume 2—Leadership* (Maxwell Air Force Base, AL: Curtis E. LeMay Center for Doctrine Development and Education, August 2015), 31, available at <https://doctrine.af.mil/dnv1vol2.htm>.

[3] Lord Moran, *The Anatomy of Courage* (London: Constable, 1945), 192.

[4] Ibid., 201, 207-208.

[5] U.S. Code, Title 10—Armed Forces, Sections 5947, 3583, and 85831, Requirement for Exemplary Conduct, as quoted in Cornell University Law School, Legal Information institute. Available at <www.law.cornell.edu/uscode/text/10>.

[6] Harold R. Winton, *Corps Commanders of the Bulge: Six American Generals and Victory in the Ardennes* (Lawrence: University Press of Kansas, 2007), 60.

[7] Eugene B. Sledge, *With the Old Breed at Peleliu and Okinawa* (New York: Presidio Press, 1981; Oxford: Oxford University Press, 1990), 40.

[8] *Marine Corps Manual*, Paragraph 1100, quoted in appendices, MCWP 6-11, 97.

[9] William E. DePuy, "11 Men 1 Mind," *Army* 8, no. 8 (March 1958), 22-24; 54-60. Reprinted in *Selected Papers of General William E. DePuy*, compiled and edited by Richard Swain, Donald L. Gilmore, and Carolyn D. Conway (Fort Leavenworth, KS: Combat Studies Institute, 1994), 17.

[10] *The Armed Forces Officer* (Washington, DC: U.S. Government Printing Office, 1960), ii.

[11] *The Armed Forces Officer* (Washington, DC: U.S. Government Printing Office, 1950), 87-88.

[12] Much of the important work on generational expectations has been done by Leonard Wong at the U.S. Army War College's Strategic Studies Institute. A number of easily accessed publications on the topic are available at <www.strategicstudiesinstitute.army.mil/pubs/people.cfm?authorID=1>.

[13] John Masters, *Road Past Mandalay: A Personal Narrative* (New York: Bantam Books, 1979), 325-326. Masters's book is a study of command under difficult conditions.

[14] A history of Army leadership doctrine can be found in an Antioch University dissertation submitted by a retired Army officer. See James Michael Young, "To Transform a Culture: The Rise and Fall of the U.S. Army Organizational Effectiveness Program, 1970–1985," available at <http://aura.antioch.edu/etds/158>. See also appendix A: "US Army Doctrinal Definitions of Leadership," and appendix B: "A Critical Analysis and Assessment of US Army Leadership Doctrine, 1946–2006." Young's work, along with the series of Army leadership manuals, can be found at<http:armyoe.com/Page_5html>.

[15] Army Doctrine Publication (ADP) 6-22, *Army Leadership* (Washington, DC: Headquarters Department of the Army, August 2012), ii.

[16] Ibid., 1.

[17] Ibid.

[18] Ibid., foreword.

[19] At the same time, published Army doctrine on leadership is no longer limited to the booklets of that name alone but, understood in accord with the Chief of Staff's encapsulation of being a leader, is to be found in a family of handbooks, among which are such titles such as ADP and Army Doctrine Reference Publication (ADRP) 6-22; ADP 1, *The Army* (Washington, DC: Headquarters Department of the Army, September 2012); ADRP 1, *The Army Profession* (Washington, DC: Headquarters Department of the Army, June 2015); and ADP and ADRP 6.0, *Mission Command* (Washington, DC: Headquarters Department of the Army, May2012). The current Field Manual (FM) 6-22, *Leader Development* (Washington, DC: Headquarters Department of the Army, June 2015), supersedes a 2006 edition titled *Army Leadership* that was formally designated "the Army's keystone field manual on leadership." The current version of FM 6-22 has been retitled and focused more narrowly on development rather than practice. Reference to it as a "keystone" has been eliminated.

[20] ADP 6-22, figure 1, "Underlying Logic of Army leadership," iii.

[21] MCWP 6-11, 30.

[22] General John Lejeune quoted in ibid., 35.

[23] *Marine Corps Manual*, W/CHG 1-3 (Washington, DC: Headquarters Department of the Navy, 1980), 1-23.

[24] "The Navy Leader Development Strategy," available at <www.usnwc.edu/getattachment/4b847467-0340-4b21-b0a2-1b071f213e34/NLDS-TheStrategy.aspx>. Walter E. Carter, Jr., "President's Forum," *The Naval War College Review* 67, no. 1 (Winter 2014), 13–15, available at <www.usnwc.edu/getattachment/4e57cdf8-f82b-4397-8b6a-22dde55949e8/President-s-Forum.aspx>.

[25] U.S. Naval War College Public Affairs, "New Center Imparts Ethical Command Leader Development," Story Number: NNS140506-13 (May 6, 2014), available at <www.navy.mil/submit/display.asp?story_id=80824>.

[26] *Leadership—Volume 2.*

[27] Air Force Manual 36-2643, *Air Force Mentoring Program* (Washington, DC: Headquarters Department of the Air Force, May 1, 2013).

[28] Mike Mullen, "The Leadership 'Playbook,'" *Surface Warfare* (January/February 1999), 36–37.

[29] Ibid., 37.

[30] Remarks at the Commencement Address for the U.S. Naval Academy, as Delivered by Admiral Mike Mullen, Annapolis, Maryland, May 23, 2008. Formerly listed on Department of Defense Speeches page at <http://www.jcs.mil/speech.aspx?ID=1078>. Currently available on You Tube as photo recording.

[31] Sir John Winthrop Hackett, *The Profession of Arms* (New York: Macmillan, 1983), 216.

[32] General Martin E. Dempsey, speech delivered to the West Point Class of 2013 500th Night, West Point, NY, January 21, 2012, available at at <www.jcs.mil/Media/Speeches/tabid/3890/Article/571855/gen-dempseys-remarks-at-the-west-point-class-of-2013-500th-night.aspx>.

[33] Major General James G. Harbord, quoted in *The Armed Forces Officer* (1950), 159.

[34] Frederick M. Franks, Jr., "Battle Command: A Commander's Perspective," *Military Review* 76, no. 3 (May–June 1996), 4, available at <http://cgsc.contentdm.oclc.org/cdm/ref/collection/p124201coll1/id/438>.

[35] *The Armed Forces Officer* (1950), 151.

[36] *Newsweek*, April 18, 1988, as found in Peter G. Tsouras, *Warriors' Word—A Quotation Book* (London: Arms and Armor Press, 1992), 302.

[37] John Baynes, *Morale: A Study of Men and Courage* (Garden Park, NY: Avery, 1988), 186–187.

[38] *The Armed Forces Officer* (1950), 141.

[39] James N. Mattis, "Ethical Challenges in Contemporary Conflict: The Afghanistan and Iraq Cases," The 2004 William C. Stutt Ethics Lecture sponsored by The Center for the Study of Professional Military Ethics, U.S. Naval Academy, November 2004, 10.

[40] Colin Powell, "1998 Sylvanus Thayer Award Speech," *Assembly* (November/December 1998), 79, available at <www.westpointaog.org/page.aspx?pid=496>.

[41] Thomas B. Buell, "Spruance Hall Dedication Address," *Naval War College Review* 25, no. 4 (March–April 1973), 18, available at <www.usnwc.edu/Publications/Naval-War-College-Review/Press-Review-Past-Issues.aspx>.

[42] For General Keene's own account, see <http://conversationswithbillkristol.org/video/jack-keane/?start=3327&end=4153>.

[43] Jim Garamone, "Pace Passes Along Combat Lessons to West Point Cadets," American Forces Press Service, April 28, 2005, available at <http://osd.dtic.mil/news/Apr2005/20050428_803.html>. Text completed using transcript of session: Peter Pace, Extemporaneous Remarks as delivered to the Constitution and Military Law Course, West Point, NY, April 27, 2005. Transcript in possession of the authors.

[44] Peter Pace, quoted in Robert Burns, "Pace Says He Refused to Quit Voluntarily," Associated Press, available at <www.military.com/NewsContent/0,13319,139244,00.html>.

[45] *The Armed Forces Officer* (1950), 159.

[46] Ibid., 160.

[47] Sir William Slim, quoted in Masters, *Road Past Mandalay*, 39–40.

[48] S.L.A. Marshall, *Men Against Fire: The Problem of Battle Command* (Norman: University of Oklahoma Press, 1947, 2000).

[49] *The Armed Forces Officer* (1950), 148–149.

[50] Ibid., 155–156.

CHAPTER SIX

The Officer at Work: Command

14. Command is the authority which an individual in the military service lawfully exercises over subordinates by virtue of rank or assignment.

15. Command and leadership are inseparable. The qualities of leadership are indispensable to a commander. Whether the command be large or small and whether the exercise of the functions of command be complex or simple, the commander must be the controlling head, his must be the master mind, and from him must flow the energy and the impulse which are to animate all under him.

16. In the practice of his task, the commander must keep in close touch with all subordinate units by means of personal visits and observation; it is essential that he know from personal contact the mental, moral, and physical state of his troops, the conditions with which they are confronted, their accomplishments, their desires, their needs, and their views, and that he promptly extend recognition for services well done, extend help where help is needed and give encouragement in adversity, but never hesitate to exact whatever effort is necessary to attain the desired end. Considerate and devoted to those whom he commands, he should be faithful and loyal to those who command him.

—War Department
Field Service Regulations, United States Army, 1923

Command is the acme of military leadership, the goal toward which officers most often aspire, and the route to the highest positions of trust in the profession of arms. Command is "the authority that a commander in the armed forces lawfully exercises over subordinates by virtue of rank or assignment."[1] Commanders at every echelon have a unique responsibility to *make sense* of the situation in which they find their forces and take all necessary actions to achieve their superiors' assigned or implicit ends. Commanders are uniquely empowered to enforce their orders and those issued under their authority. They retain comprehensive responsibility for the conduct, efficiency, effectiveness, and health and welfare of all the forces entrusted to them. Though seldom mentioned explicitly today, commanders are still expected, as the 1923 Army Field Service Regulation required, "never [to] hesitate to exact whatever effort is necessary to attain the desired end."[2] The French historian Marc Bloch, who was a World War I infantry officer, World War II staff officer, and Resistance martyr, wrote about the determination, even ruthlessness, required in adversity, in his stinging critique of French defeatism in 1940:

> *What, probably, more than anything else marks the true leader is the power to clench his teeth and hang on, the ability to impart to others a confidence that he feels himself. . . . Above all, he must be willing to accept for the men under him, no less than for himself, sacrifices which may be productive of good, rather than a shameful yielding which must remain forever useless.*[3]

Command is held only by virtue of appointment or, temporarily, by succession in cases of unexpected vacating of office by an incumbent, either by relief or incapacitation. There is some ambiguity in the Armed Forces over the issue of who may command. The third definition of Section 801 of Title 10, U.S. Code, the opening section of the Uniform Code of Military Justice (UCMJ), says that "the term 'commanding officer' includes only commissioned officers," while the *Marine Corps Manual*, under "Eligibility to Command," states that "any commissioned, warrant, or noncommissioned officer of the Marine Corps is eligible to command activities of the Marine Corps subject to limitations imposed by the Commandant of the Marine Corps or higher authority."[4] Army Regulation (AR) 600-20, *Army Command Policy*, following Section 801, states:

> *A commander is . . . a commissioned or WO* [warrant officer] *who, by virtue of grade and assignment, exercises primary command authority over a military organization or prescribed territorial area that under pertinent official directives is recognized as a 'command.'... A civilian, other than the President as Commander-in-Chief (or National Command Authority), may not exercise command.*[5]

The Secretary of Defense, placed in the chain of command by the 1986 Department of Defense Reorganization Act, is included within the term National Command Authority.

The exercise of command involves visualizing a future state, normally within the intentions of a superior commander, planning and directing the activities of the subordinate organization to achieve that state, following through to ensure and harmonize performance, rewarding good performance, and correcting or sanctioning bad. The *Marine Corps Manual* lists the following as the inherent attributes of command:

1. precedence over all persons commanded
2. power to enforce the official will of the commander through the issuance of necessary directives
3. authority to make inspections to ensure compliance with such directives
4. authority to initiate or apply authorized disciplinary measures.[6]

Command, like other forms of leadership, involves human as well as legal relationships, and therefore relies on character as much as formal authority for its effectiveness. In 2011, Chief of Naval Operations Admiral Gary Roughead wrote to prospective Navy commanding officers that "a Commanding Officer must possess professional competence, intelligent good sense, the 'nicest sense of personal honor' and meet our highest standards of personal conduct and leadership."[7] Earlier, Roughead reminded command selectees that:

> *As a Commanding Officer, you must build trust with those Officers and Sailors under your command. You build trust through your character and in your actions which demonstrate professional competence, judgment, good sense, and respect for those you lead. This trust can only be built through personal interaction on a daily basis at every level in your chain-of-command. Human interaction remains the dominant factor in leading Sailors.*[8]

In the Armed Forces, command is exercised within a chain of command, a web of appointed commanders with the President at the top and the lowest privates, seamen, or airmen at the bottom, and with parallel and overlapping responsibilities necessary to ensure effectiveness within a large and decentralized organization. This chapter addresses specifically the purpose of the chain of command; notions of individual authority, responsibility, and accountability shared by the Armed Forces; and the character attributes expected of Armed Forces officers in command.

Chain of Command

Large forces are articulated by chains of command for purposes of flexibility and to accommodate limits on span of control. The forces of the overall commander are divided among subordinate commanders in accordance with the superior's vision of operations. Each subordinate commander is given a grant of authority, and assigned his or her own responsibilities within the scope of the superior's, yet the superior retains full responsibility over all functions of the whole force.

In any large force, tension exists between the senior commander's comprehensive responsibility and the need to decentralize action. Local commanders must have the ability to exercise initiative to adapt their actions to local conditions and immediate contingencies, within the superior commander's intentions and without disrupting the coherence of the entire force of which subordinate units are only a part. Senior commanders establish standards for their subordinates and inspect periodically to ensure their maintenance. The goal, according to Admiral Ernest King, is that "each does his own work in his own sphere of action or field of activity."[9]

The proper balance between decentralized execution and comprehensive action under centralized responsibility is a perennial concern. In 1941, Admiral King was worried that over-centralization, manifested in detailed instructions, would sap the exercise of initiative by local commanders when the U.S. Navy entered the war. He sent out two memorandums, in January and April 1941, to address the issue. In the first, he emphasized the importance of senior commanders freeing subordinates from restrictive orders. In the second, King addressed the obligation of subordinates to exercise their initiative within the framework of the higher commander's intentions and as "a correlated part of a connected whole."[10]

King did not settle the issue for all time. Following the war in Vietnam, prompted by civilian defense critics, the Marine Corps and the Army spent a good deal of time debating the necessity for what became known as *mission orders*, instructions issued with expectation of the exercise of individual initiative and adaptation in execution. Both services adopted the principle of intelligent obedience as the standard method of command. The Army emphasized use of initiative within the commander's expressed purpose or "intent."[11] The Marine Corps combined the expectation of "leaders with a penchant for boldness and initiative down to the lowest levels" with what it called "Mission Tactics."[12] Notably, while both services attempted to get away from what was perceived to be a "zero defects" mentality to allow for a degree of risk-taking, the Marine Corps, like King before, warned that initiative was not to be understood as license. "It does not mean that commanders do not counsel subordinates on mistakes; constructive criticism is an important element of learning. Nor does it give subordinates free license to act stupidly or recklessly."[13]

In August 2003, the Army published what remains its most thorough doctrinal investigation into the concept of command, Field Manual (FM) 6-0, *Mission Command: Command and Control of Army Forces*. The primary focus of the manual was to deconstruct the then-overarching concept of command and control into the individual practice of command (*Command*), characterized as an art, from the technical and organizational systems, characterized as a science, created to carry it out (*Control*).[14] Within this construct, the manual recognized two archetypes of command: directive command and mission command. FM 6-0 adopted explicitly a preference for *mission command*, which it defined concisely as "the conduct of military operations through decentralized execution based upon mission orders for effective mission accomplishment."[15]

More recently, based on his conclusions drawn from the wars in Afghanistan and Iraq, General Martin Dempsey, first as commander of U.S. Army Training and Doctrine Command (2008), then as Chief of Staff of the Army (2011), and later as Chairman of the Joint Chiefs of Staff (2011–2015), adopted mission command as a personal signature issue. The month Dempsey left the office of Army Chief of Staff to become Chairman of the Joint Chiefs, the Army published the successor volume to the 2003 FM 6-0 titled simply, *Mission Command* (September 13, 2011), signed by Dempsey's successor, General Raymond T. Odierno. This manual restructured the 2003 concept by expanding the notion of Mission Command, *as a philosophy,* to comprehend the entire function formerly called Command and Control and divided it into an art or philosophy of command, and what it called "the mission command warfighting function."[16] Mission Command was now defined (as an Army term) as the "exercise of authority and direction by the commander using mission orders to enable disciplined initiative within the commander's intent to empower agile and adaptive leaders in the conduct of full spectrum operations."[17]

At the same time, joint doctrine writers adopted the Mission Command terminology as well. In August 2011, before the retirement of Admiral Mike Mullen as Chairman of the Joint Chiefs of Staff, the Joint Staff published Joint Publication (JP) 3-0, *Joint Operations*, which defines Mission Command as "the conduct of military operations through decentralized execution based upon mission-type orders." The joint manual goes on to say that "successful mission command demands that subordinate leaders at all echelons exercise disciplined initiative and act aggressively and independently to accomplish the mission."[18] General Dempsey, who succeeded Mullen in September, subsequently issued a White Paper on Mission Command that supplemented and extended the August 2011 JP 3-0 discussion.[19] All of this history seems to suggest that achieving the proper vertical balance between centralized responsibility and decentralized execution can be expected to remain a matter of continuous adjustment, but that certain principles involving the balance between authority and responsibility remain lasting.

Authority—Responsibility—Accountability

The concepts of authority, responsibility, and accountability are associated inextricably with one another within the idea of command. It is all but impossible to speak of one without reference to the other two. Since the ancient Greeks, the idea of *responsibility* has meant that an individual or collectivity is expected to perform some duty in a satisfactory manner based upon commonly accepted standards.[20] The most common evidence that such an expectation exists is the anticipation of some sanction (accountability) in the event of failure or omission. A necessary prior condition is that the individual said to be responsible has the power and ability to do what is expected on the basis of some recognized authority. Absent authority and accountability, it is difficult to see how responsibility can be said to attach.

Authority. Today, one premise on which all the Services agree is the notion that in order to be effective in accomplishing assigned responsibilities, there must be a corresponding grant of authority and necessary freedom of action. The separate Services all agree in principle on the comprehensiveness and irreducibility of a commander's responsibility. A commander's authority is derived in the first instance from grants of power in law, Department of Defense Directives, and Service Regulations. The Uniform Code of Military Justice underwrites the chain of command and assigns important procedural roles to senior commanders. The legal and regulatory authorities vested in commanders generally are further enhanced by specific powers granted under the authority of immediate commanders.

Responsibility. The Air Force leadership manual follows joint doctrine when it says: "Command includes the authority and responsibility for effectively using available resources and for planning the employment of, organizing, directing, coordinating, and controlling military forces for the accomplishment of assigned missions. It also includes responsibility for health, welfare, morale, and discipline of assigned personnel."[21] The Army Command Regulation charges its commanders with promoting a positive environment, developing in Soldiers a sense of duty, defined as "obedient and disciplined performance"; integrity; and respect for their authority. The last they are to do by developing "the full range of human potential in their organization," informing troops of the need for military discipline, and "properly training their Soldiers and ensuring that both Soldiers and equipment are in the proper state of readiness at all times."[22]

Because of the unique character of the responsibility of command at sea, it is the Navy that has traditionally emphasized the greatest authority in command. This authority responds to the conditions under which command at sea occurs and the unitary responsibility of the ship's captain for both the security of the vessel and the welfare of its Sailors. Because warships operate in a hostile environment, and very often independently, distant from close oversight, ship captains have traditionally enjoyed significant authority and independence of action while underway.[23] The English author Joseph Conrad wrote of the ship captain's unique responsibility for the welfare of his ship, describing the observations of an officer of the watch on a merchant ship whose captain comes on deck during a violent storm:

> *Jukes was uncritically glad to have his captain at hand. It relieved him as though that man had, by simply coming on deck, taken most of the gale's weight upon his shoulders. Such is the prestige, the privilege, and the burden of command.*
>
> *Captain MacWhirr could expect no relief of that sort from anyone on earth. Such is the loneliness of command.*[24]

Former Coast Guard Commandant Thad Allen notes another aspect of the loneliness of command (or any senior leadership position):

> *you've got to learn how to manage your own morale. When you're in a situation like many commanding officers are in or people that are running large complex responses, there are not a whole lot of people around that can buoy your spirits, give you positive feedback. There are going to be a lot of times where you're going to get negative feedback for a long, long time before you get any positive feedback. You have to be able to ascertain what you need to do, lay out a course of action, identify the effects to be achieved, and then go after that; and you have to do that with a fairly stable emotional base to work from. That's not easy to do because you can get very angry and frustrated.*[25]

The U.S. Navy's submarine service offers an iconic example of the captain's responsibility for his ship in the final actions of Commander Howard W. Gilmore, the skipper of the submarine U.S.S. *Growler* in February 1943. Mortally wounded though still conscious, lying outside on the deck of his boat during a surface fight with a Japanese gunboat, Gilmore gave a final decisive order: "*Take her down*," he stated, ending his own life but saving his boat and its crew.[26] Gilmore's gallantry and intrepidity were recognized by posthumous award of the Congressional Medal of Honor and a memorial plaque and room at the U.S. Naval Academy in Annapolis.

In the early 1970s, as the Vietnam War ran down and serious acts of indiscipline were reported in U.S. military forces, Admiral Arleigh Burke addressed an audience at the Naval War College on "The Art of Command." Known as "31-knot Burke," the admiral was a famous destroyer commander during World War II. He participated in the United Nations delegation to the initial peace talks in Korea (led by U.S. Vice Admiral C. Turner Joy) and, while only a rear admiral, was selected to be Chief of Naval Operations by President Dwight Eisenhower. "Every man in the military service spends his whole time in the service seeking to improve his role in the command system," Burke told his audience, "both by being ready to carry out in an effective manner all orders he may receive and by being capable and willing to give orders to his unit to further the operation his outfit is undertaking."[27]

Burke went on to argue for the importance of matching responsibility with authority and expressed concern that local authority was being undermined by over-supervision from on high. At the same time, he recognized the need for higher-level commanders to maintain control by establishing and demanding adherence to strict standards. "The most important responsibility of every commander," he said, was "the responsibility to insure that the standards he sets for his unit are high enough to enable his unit to be successful and, as a corollary, to reward those subordinates who do extraordinarily well and to punish those who fail." He was critical of a force he thought too willing to reward people and too hesitant "to punish those who fail to measure up to high standards."[28]

Commanders are responsible for speaking truth to authority. In fact, senior commanders depend upon frankness from those responsible to them for execution of their orders. In his memoir, General Matthew Ridgway addressed the responsibility of a commander to identify and oppose bad ideas that will lead to unnecessary, or at least improvident, losses to his Soldiers. Discussing a scheme to drop his 82nd Airborne Division on Rome in an attempt at a *coup de main*, Ridgway relates that he went all the way to the Allied Theater Commander, then General Sir Harold Alexander, to express his concerns. With help from Alexander's Chief of Staff, Walter Bedell Smith, he was able to convince Alexander to send a two-man reconnaissance team to meet with the Italian government, which was supposed to be prepared to assist in the landing. Maxwell Taylor, Ridgway's artillery commander, led the party. He retired in 1959 but was recalled by President Kennedy to serve as Military Advisor to the President, then Chairman of the Joint Chiefs of Staff, and subsequently as ambassador to South Vietnam.

Taylor reported back by radio that conditions were not propitious, and the mission was canceled with Soldiers and planes on the runway. Ridgway wrote:

> *It seems to me, too, that the hard decisions are not the ones you make in the heat of battle. Far harder to make are those involved in speaking your mind about some hare-brained scheme which proposes to commit troops to action under conditions where failure is almost certain, and the only results will be the needless sacrifice of priceless lives. When all is said and done, the most precious asset any nation has is its youth, and for a battle commander ever to condone the needless sacrifice of his men is absolutely inexcusable. In any action you must balance the inevitable cost in lives against the objectives you seek to attain. Unless, beyond any reasonable doubt, the results reasonably to be expected can justify the estimated loss of life the action involves, then for my part I want none of it.*[29]

Easily lost sight of is that, in addition to aggressively opposing the mission, Ridgway first offered a useful alternative (sending Taylor to Rome) to mitigate the risk. Then, having apparently failed to convince his superiors of the futility of the effort, Ridgway was prepared to lead his troops in the attempt and give his best efforts to make it succeed. Coincidentally, in a 1920 letter to a retired friend at the Virginia Military Institute, George C. Marshall had written that an officer should "make a point of extreme loyalty, in thought and deed, to your chiefs personally; and in your efforts to carry out their plans or policies, the less you approve the more energy you must direct to their accomplishment."[30]

As key members of the profession of arms, commanders are responsible for the professional development of their subordinates, particularly subordinate officers, for success in positions in leadership. This calls for observing subordinates' state of individual training, correcting them when they make mistakes, and seeing that they are retrained to the necessary standards when that is required. Ultimately the commander is called upon to distinguish the successful from the unsuccessful so the institution can reward those most capable and remove those less so.

As the senior representative of the profession in any unit, the commander has a special responsibility both to model the behaviors valued by the profession and to encourage subordinates in their emulation. The commander must set the example, and create an ethical space within which collective reflection on the military calling is frequent, accepted, and instructive. Normally this requires the human touch, an ability to engage informally with subordinates, and to coach and mentor without creating a sense of unease with those whose professional futures are very much in the commander's hands. The truly gifted commanders can have a life-long influence on the profession by this action alone.

Accountability. Like responsibility, accountability is not limited to commanders, but extends to all leaders in authority, indeed all members of the Armed Forces. Accountability involves accepting the consequences for the outcomes of action or inaction in circumstances for which one bears responsibility—whether it results from individual action, inaction, or inattention. Accountability may result in judicial or administrative sanction. The professional grant of discretionary authority enjoyed by Armed Forces officers, and especially commissioned officers, does not extend to violation of law, even for very senior commanders otherwise granted wide latitude. Officers can be relieved of their offices for misconduct, and retired at a lower grade, even when no judicial action is called for.

In all the Services, the priority necessarily given judicial action delays and obscures the importance of the administrative sanction in assignment of accountability. In 1995, General Ronald Fogleman, the Air Force Chief of Staff, distributed a video tape to the Air Force titled "Air Force Standards and Accountability." In it, he explained administrative action he had directed against Air Force personnel deemed responsible for a 1994 incident in which Air Force aircraft shot down two Army Black Hawk helicopters in Northern Iraq. The incident gained immediate notoriety, as incidents of fratricide can do, and it took some time for the normal processes to arrive at the legal determination of what action was called for under the Uniform Code of Military Justice.

In the video, the Chief of Staff said that the Secretary of the Air Force had directed him to review all actions taken subsequent to judicial inquiry, including "adequacy of evaluations, decorations, subsequent assignments, promotions and retirements." One major concern was "that Air Force standards be clearly understood [as well as] the necessity that individuals be held accountable for meeting those standards," even where punitive legal action was not called for. Observing that the judicial process in the Black Hawk case had produced no criminal prosecutions or convictions, the Chief declared that "Air Force standards require far more than mere compliance with the law. They require that people display the extraordinary discipline, judgment and training that their duties require and the American people expect."[31]

General Fogleman's review had disclosed a number of inconsistent post-incident administrative actions, particularly with regard to performance evaluations that did not reflect the seriousness of the incident in which 26 friendly Soldiers, Airmen, and civilians lost their lives. As a result, the Chief of Staff issued supplemental performance evaluations and removed the responsible officers from flight status for a minimum of 3 years. The Chief admonished all Air Force rating officers, stating that

> *It is important for commanders and raters to remember that your ratings, comments and actions do not represent arbitrary action against the individual, but reflect an appropriate response*

to their misconduct or failure to meet standards. And recognize that your loyalty and commitment must be to the larger organization—to the Air Force as an institution. Air Force standards must be uniformly known, consistently applied and non-selectively enforced. Accountability is critically important to good order and discipline of the force. And, failure to ensure accountability will destroy the trust of the American public.[32]

In the naval Services, the unique responsibility of command at sea is coupled with what to other services often seems a kind of draconian accountability. In the words of Commander Roger D. Scott:

The doctrine of command accountability is most strictly applied to command at sea in recognition of the fact that naval vessels frequently operate independently, far from sources of assistance, in an environment made hostile by the elements or by enemies. Life at sea is surrounded by dangerous forces on the ship and around it. Mistakes and omissions can mean death of all hands on board.[33]

Naval officers can be, and often are, relieved of command for no more stated reason than "loss of confidence in an officer in command."[34] Even though other services adopt similar formulations, the Navy often seems more rigorous in its application. Subsequently, Scott wrote: "The traditional scope of duties and accountability that attach to command at sea [have] no parallel in the military or civilian spheres."[35]

The doctrine of command accountability in the Navy is enshrined in paragraph 0802 of Navy Regulations. "The responsibility of the commanding officer for his or her command is absolute. . . ."[36] In 1991, Captain Larry Seaquist, USN, a prior captain of the battleship *Iowa*, wrote to the *Navy Times* on the occasion of the publicity and discussion of the gun explosion on the *Iowa*, which killed all those manning a main gun turret:

Accountability is a severe standard: The commander is held responsible for everything that occurs under his command. Traditionally, the only escape clause was "an act of God," an incident that no prudent commander could reasonably have

> *foreseen. And "reasonably" was tied to the requirement to be "forehanded"—a Sailor's term dictating that even unlikely contingencies must be thought through and prepared for. The penalties of accountable failure can be drastic: command and career cut short, sometimes by court-martial.*[37]

Seaquist's article echoed a 1952 *Wall Street Journal* editorial addressing an inquiry into the sinking of the destroyer-minesweeper USS *Hobson* by the carrier USS *Essex* in a collision at sea in which 176 Sailors were lost:

> *It is cruel this accountability of good and well-intentioned men. But the choice is that or an end to responsibility and finally, as the cruel sea has taught, an end to the confidence and trust in the men who lead, for men will not long trust leaders who feel themselves beyond accountability for what they do . . . when men lose confidence and trust in those who lead, order disintegrates into chaos and purposeful ships into floating derelicts.*[38]

In keeping with the democratic foundation of the United States, the UCMJ makes the actions of any commander subject to superior review in cases where subordinates feel they have been wronged. Any military subordinate may file a formal request for redress under Article 138 (Section 938 of Title 10). Such a request must be forwarded for resolution to the officer exercising general court-martial jurisdiction over the commander in question. He or she must report to the Service secretary the action taken to resolve the issue. Additionally, officers, including senior commanders possessing wide latitude of discretion otherwise, are held to strict standards for financial propriety, as in use of government transportation and submission of claims for reimbursement for official travel, and for observance of the financial strictures that Congress imposes as part of their Constitutional role of executive oversight. More than one commander, with an otherwise extraordinary record, has stumbled on such limits, when an aggressive "can do attitude," and a bit of hubris and impatience with fiscal regulation, run into legal restrictions that seem unduly confining in view of the good anticipated from the deviation taken.

Commanders possess authority to charge subordinates with criminal offenses under the UCMJ, convene military courts-martial, and review findings and sentences as elements of their command authority. Recently, however, the extent of senior commanders' review authority, under Article 60 of the Uniform Code, has been reduced significantly in light of perceived command failures in enforcing sexual misconduct policies.[39] Two Air Force general officer commanders, one female and one male, were denied promotion and continued service by Congress for failing to uphold courts-martial decisions in cases of sexual assault.[40] Both officers acted within their existing authorities. They were sanctioned for what members of the Senate believed was bad faith or bad judgment, and consequently the professional leadership of the Armed Forces lost a measure of its authority over administration of the system of military justice through congressionally driven changes in the UCMJ. These actions are indicative of the inherent subjugation of commanders to individual accountability for the execution of their offices. These incidents also demonstrate the divided authority between the President's authority of appointment and Congress's ability to enforce standards under its Constitutional authority to raise and support armies and "to make Rules for the Government and Regulation of the land and naval forces," and the Senate's authority to confirm general officer appointments.

Character

Command of ground forces is, for the most part, less independent than command at sea, precisely because senior officers can visit subordinates and observe the state of the command with some regularity. At least theoretically, the command of ground forces is as encompassing as that at sea: the commander is responsible for everything the command (or its members) does or fails to do. The character of ground combat commanders too is a subject of frequent comment.

In the earliest versions of *The Armed Forces Officer*, one of S.L.A. Marshall's strongest chapters addressed *Esprit.* Marshall believed that the commander's authority stemmed largely from the soldiers' perception of his character. He further argued that

> *the custodianship of* esprit *must ever be in the hands of the officer corps. When the heart of the organization is sound, officership is able to see its own reflection in the eyes of the enlisted man . . . insofar as his ability to* [mold] *the character of troops is concerned, the qualifying test of the leader is the judgment placed upon his military abilities by those who serve under him. If they do not deem him fit to command, he cannot train them to obey.*[41]

The source of their approval was not to be won simply by courageous acts. Troops, Marshall wrote, "can be kept in line under conditions of increasing stress and mounting hardship, only when loyalty is based upon a respect . . . won by consistently thoughtful regard for the welfare and rights of his men, and a correct measuring of his responsibility to them."[42]

World War II provided numerous examples of company commanders who, through strength of character, established emotional ties with their men. One was Captain Henry T. Waskow of Belton, Texas, immortalized by Ernie Pyle in a wartime column and portrayed by Robert Mitchum in the postwar movie, *G.I. Joe*. Waskow, killed by a mortar round in Italy, "had led his company since long before it left the States," wrote Pyle. "He was very young, only in his middle twenties, but he carried in him a sincerity and a gentleness that made people want to be guided by him."[43]

Eugene B. Sledge's memory of his company commander was quoted in chapter 5. Sledge reflected further on the impact of Captain Haldane's death:

> *Our company commander represented stability and direction in a world of violence, death, and destruction. . . . We felt forlorn and lost . . . he commanded our individual destinies under the most trying conditions with the utmost compassion . . . the loss of our company commander at Peleliu was like losing a parent we depended upon for security—not our physical security, because we knew that was a commodity beyond our reach in combat, but our mental security.*[44]

Leadership by more senior commanders is less intimate. Higher-level commanders lack the personal relationship to troops that regimental officers and division chiefs enjoy. But senior commanders also derive authority from the character they exhibit.

General Matthew Ridgway gave his views on the importance of a commander's character in a speech to the Army Command and General Staff College in May 1966. He told a story about the fight on the north shoulder of the Battle of the Bulge during the German Ardennes offensive:

> *Another corps commander just entering the fight next to me remarked: "I'm glad to have you on my flank. It's character that counts." I had long known him, and I knew what he meant. I replied: "That goes for me, too." There was no amplification. None was necessary. Each knew the other would stick however great the pressure; would extend help before it was asked, if he could; and would tell the truth, seek no self-glory, and everlastingly keep his word. Such feeling breeds confidence and success.*[45]

Notes

[1] Joint Publication 1-02, *The Department of Defense Dictionary of Military and Associated Terms* (Washington, DC: The Joint Staff, November 8, 2010, as Amended Through 15 January 2015), 40, available at <www.dtic.mil/doctrine/new_pubs/jp1_02.pdf>.

[2] Extracted from Chapter II: Command and Staff of War Department, Field Service Regulations United States Army 1923 (Washington, DC: U.S. Government Printing Office, 1924), 4.

[3] Marc Bloch, *Strange Defeat: A Statement of Evidence Written in 1940* (New York: Norton, 1968), 111.

[4] *Marine Corps Manual* (Washington, DC: Headquarters Department of the Navy, 1980), section A, Organization, Functions and Command, paragraph 1006, subparagraph 3, 1-13.

[5] Army Regulation 600-20, *Army Command Policy* (Washington, DC: Headquarters Department of the Army, November 6, 2014), paragraph 1-5, 1. Section 162 of Title 10 specifies the chain of command runs from the President, through the Secretary of Defense, to combatant commanders.

[6] *Marine Corps Manual*, 1-13.

[7] U.S. Department of the Navy, Chief of Naval Operations, "Memorandum for All Prospective Commanding Officers, Subject: The Charge of Command," June 9, 2011.

[8] Ibid.

[9] Admiral Ernest King, Memorandum: From Commander-in-Chief, United States Atlantic Fleet, Subject: "Exercise of Command—Correct Use of Initiative," April 22, 1941, reprinted in "Set and Drift," *The Naval War College Review* 28, no. 3 (Winter 1976), 95, available at <www.usnwc.edu/Publications/Naval-War-College-Review/ArchivedIssues/1970s/1976-Winter.aspx>.

[10] Ibid., 96. The January memorandum is on pages 93-94 under the section titled "Set and Drift."

[11] Field Manual (FM) 100-5, *Operations* (Washington, DC: Headquarters Department of the Army, May 1986), 3-4, 15.

[12] Marine Corps Doctrinal Publication (MCDP 1) [Formerly Fleet Marine Force Manual 1], *Warfighting* (Washington, DC: Headquarters Department of the Navy, 1997), 57-58, 87-91.

[13] Ibid., 57.

[14] FM 6-0, *Mission Command: Command and Control of Army Forces* (Washington, DC: Headquarters Department of the Army, August 11, 2003), 1-1–1-4.

[15] Ibid. 1-17. The preface to the manual states that "FM 6-0 establishes mission command as the C2 concept for the Army," viii.

[16] All taken from the "Introduction" to FM 6-0, *Mission Command* (Washington, DC: Headquarters Department of the Army, September 13, 2011), in the form of a Kindle book. The Army has since replaced the 2011 version with Army Doctrine Publication and Army Doctrine Reference Publication 6-0, both dated May 17, 2012, both titled *Mission Command.* A May 2014 version of FM 6-0 is titled *Commander and Staff Organization and Operations*, and contains very different subject matter than the original 6-0 series.

[17] Ibid., "Glossary."

[18] Joint Publication 3-0, *Joint Operations* (Washington, DC: The Joint Staff, August 11, 2011), 11-2.

[19] Martin E. Dempsey, "Mission Command White Paper," April 3, 2012, 1.

[20] Andrew Eshleman, "Moral Responsibility," in *The Stanford Encyclopedia of Philosophy* (Stanford, CA: The Metaphysics Research Lab, Center for the Study of Language and Information, 2014), available at <http://plato.stanford.edu/archives/sum2014/entries/moral-responsibility>.

[21] *Leadership—Volume 2* (Maxwell Air Force Base, AL: Curtis E. LeMay Center for Doctrine Development and Education, August 2015), 10-11, available at <https://doctrine.af.mil/dnv1vol2.htm>.

[22] Army Regulation 600-20, paragraph 1-5, "Command," 2.

[23] This sentence is intentionally written under influence of the Scott quotation on following page. (note 30).

[24] Joseph Conrad, *Typhoon* (New York: G.P. Putnam's Sons, 1902), 75.

[25] Rich Cooper, "An Interview with Adm. Thad Allen (USCG-Ret.)," *Defense Media Network*, November 16, 2010, available at <www.defensemedianetwork.com/stories/an-interview-with-adm-thad-allen-uscg-ret/>.

[26] Edwyn Gray, *Captains of War: They Fought Beneath the Sea* (London: Leo Cooper, 1988), 154 et seq.

[27] Arleigh A. Burke, "The Art of Command: A Lecture Delivered by Admiral Arleigh A. Burke, U.S. Navy (Ret.)," *The Naval War College Review* 24, no. 10 (June 1972), 25, available at <www.usnwc.edu/Publications/Naval-War-College-Review/ArchivedIssues/1970s/1972-June.aspx>.

[28] Ibid., 27.

[29] Matthew B. Ridgway, as told to Harold Martin, *Soldier: The Memoirs of Matthew B. Ridgway* (New York: Harper Brothers, 1956), 81-83.

[30] George C. Marshall, Letter to Brigadier General John S. Mallory, November 5, 1920, in *The Papers of George Catlett Marshall*, Vol. I, *"The Soldierly Spirit" December 1880–June 1939*, ed. Larry I. Bland and Sharon R. Ritenour (Baltimore, MD: Johns Hopkins University Press, 1981), 202.

[31] Ronald R. Fogleman, "Air Force Standards and Accountability," text of a videotape on the topic of Air Force Standards and Accountability produced following administrative actions taken against officers involved in last year's [1994] shootdown of two U.S. Army helicopters. Date of Video: August 10, 1995. Page 2 of 4. Available at <www.au.af.mil/au/awc/awcgate/readings/air_force_standards_and_acc.htm>.

[32] Ibid., 4.

[33] Roger D. Scott, "Kimmel, Short, McVay: Case Studies in Executive Authority, Law, and the Individual Rights of Military Commanders," *Military Law Review*, vol. 156 (June 1998), 169-170, available at <www.loc.gov/rr/frd/Military_Law/Military_Law_Review/pdf-files/277C79~1.pdf>.

[34] Ibid., 72-73. In the Army, the general policy statement is, "When a higher ranking commander loses confidence in a subordinate commander's ability to command due to misconduct, poor judgment, the subordinate's inability to complete assigned duties, or for other similar reasons, the higher ranking commander has authority to relieve the subordinate commander." However, in practice, only general officers (or "frocked" colonels) may relieve a subordinate without first obtaining written approval of a general officer in the chain of command. Paragraph 2-17, "Relief for cause," Army Regulation 600-20, *Army Command Policy with Rapid Action Revision (RAF) Issue Date: 20 September 2012*, 17.

[35] Scott, 168.

[36] *U.S. Navy Regulations 1990* (Washington, DC: Headquarters Department of the Navy, 1990), 47, available at <http://doni.documentservices.dla.mil/US%20Navy%20Regulations/Chapter%208%20-%20The%20Commanding%20Officer.pdf>.

[37] Larry Seaquist, "Iron Principle of Accountability Was Lost in Iowa Probe," *Navy Times*, December 9, 1991, quoted in Scott, "Kimmel, Short, McVay," note 548, 195.

[38] See also the editorial addressing the Navy inquiry into the 1952 sinking of the USS *Hobson* by collision with the *Essex* in "Hobson's Choice," *Wall Street Journal*, May 14, 1952, available at <www.thecommandingofficer.com/charge-of-command/hobson's-choice>.

[39] David Vergun, "New Law Brings Changes to Uniform Code of Military Justice," Army News Service, January 8, 2014, available at <www.defense.gov/news/newsarticle.aspx?id=121444>. See also Claire McCaskill, "An Evidence-Based Approach to Military Justice Reform," *Time*, March 15, 2014, available at <http://time.com/26081/claire-mccaskill-military-sexual-assault-bill/>.

[40] Craig Whitlock, "National Security: General's Promotion Blocked over Her Dismissal of Sex-assault Verdict," *Washington Post*, May 6, 2013, available at <www.washingtonpost.com/world/national-security/generals-promotion-blocked-over-her-dismissal-of-sex-assault-verdict/2013/05/06/ef853f8c-b64c-11e2-bd07-b6e0e6152528_story.html>. One of the officers, Lieutenant General Susan J. Helms, a former astronaut, was denied confirmation for appointment to command Air Force Space command by Senator Claire McCaskill. See also Robert Draper, "The Military's Rough Justice on Sexual Assault," *New York Times Magazine*, November 26, 2014, available at <http://www.nytimes.com/2014/11/30/magazine/the-militarys-rough-justice-on-sexual-assault.html?_r=0>. The other officer was the Commander, Third Air Force, Lieutenant General Craig Franklin, who was retired as a major general based on his reversal of a court-martial decision in Europe.

[41] *The Armed Forces Officer* (Washington, DC: U.S. Government Printing Office, 1950), 163.

[42] Ibid., 161.

[43] Ernie Pyle, *Brave Men* (New York: Henry Holt and Company, 1944), 106–107.

[44] Eugene B. Sledge, *With the Old Breed at Peleliu and Okinawa* (New York: Presidio Press, 1981; Oxford: Oxford University Press, 1990), 140–141.

[45] Matthew B. Ridgway, "Leadership," *Military Review* 46, no. 10 (October 1966), 41, available at <http://cgsc.contentdm.oclc.org/cdm/singleitem/collection/p124201coll1/id/634/rec/11>.

CHAPTER SEVEN

The Officer and Society: The Vertical Dimension

The relationship between the U.S. military profession and American society has two dimensions: the vertical, which is the domain of civilian control of the military; and the horizontal, which involves how practices and values in the military mesh—or do not mesh—with practices and values in the larger society the military is sworn to serve. Officers are engaged in both dimensions. This chapter will address the vertical dimension; the next chapter, the horizontal.

Constitutional Foundation

The military is subject to control by the three branches of the national government in accordance with their separate authorities under the Constitution. Civilian control of the military is deeply embedded in the American DNA, going back at least to the Declaration of Independence, which included as one item in its bill of particulars against King George that "He has affected to render the Military independent of and superior to the Civil Power."[1] Some 11 years later, after winning independence from Great Britain and still echoing that grievance, the drafters of the U.S. Constitution assigned every power related to the Armed Forces to civilian officials. Article I, Section 8, states that "the Congress . . . shall provide for the common Defense . . . of the United States," and further gives to the legislative branch the following important powers:

- to define and punish Piracies and Felonies committed on the high Seas, and Offences against the Law of Nations
- to declare War, grant Letters of Marque and Reprisal, and make Rules concerning Captures on Land and Water
- to raise and support Armies, but no Appropriation of Money to that Use shall be for a longer Term than two Years
- to provide and maintain a Navy
- to make Rules for the Government and Regulation of the land and naval Forces;
- to provide for calling forth the Militia to execute the Laws of the Union, suppress Insurrections and repel Invasions
- to provide for organizing, arming, and disciplining, the Militia, and for governing such Part of them as may be employed in the Service of the United States, reserving to the States respectively, the Appointment of the Officers, and the Authority of training the Militia according to the discipline prescribed by Congress.[2]

Article II, Section 2 gives other powers to the executive branch, in particular the power of command of the United States Armed Forces to the "President [who] shall be Commander in Chief of the Army and Navy of the United States, and of the Militia of the several States, when called into actual Service of the United States." Article II, Section 3, states that the President "shall Commission all the Officers of the United States."[3]

Article III establishes a Federal judicial system with a supreme court holding responsibility for review of the proceedings of inferior courts, one category of which are those created by the legislative branch to exercise military law.

The officer's commission includes an obligation of obedience, in particular to the orders of the President or the President's successors. Moreover, in accordance with Article VI of the Constitution, all "executive and judicial Officers, both of the United States and the several States, shall be bound by Oath or Affirmation to support this Constitution," an obligation which includes respect for the authorities embedded in Articles I, II, and III.

Further specifying civilian control, Title 10 of the U.S. Code establishes the chain of command for the Armed Forces of the United

States, placing two civilians in authority over all operational military commanders:

> *Chain of Command.—Unless otherwise directed by the President, the chain of command to a unified or specified combatant command runs—*
>
> *(1) from the President to the Secretary of Defense; and*
>
> *(2) from the Secretary of Defense to the commander of the combatant command.*[4]

Service secretaries within the Department of Defense exercise executive civilian control over the several military departments.

Formalizing civilian control of the military in the Constitution and Federal statutes flows from the underlying theory of democracy, namely, that the people are sovereign and exercise their authority through elected representatives and officials. Writing in the latter half of the 20th century, Morris Janowitz highlighted the underlying problem: "Analysis of the pressures of civilian control over the military leads ultimately to the full complexity of the American federal and pluralistic system of government."[5] The first words of the Constitution embody this theory: "We the People of the United States . . . do ordain and establish this Constitution for the United States of America." Those elected by the people have preeminence and authority over those in uniform, who are not chosen by the people, but rather appointed and commissioned by responsible civilian authority.

So accepted is this principle that even American popular culture makes only rare forays challenging the subordination of the military to the civilian. Perhaps the last significant example was the popular 1962 novel and 1964 film *Seven Days in May*, which appeared at the height of the Cold War. Worth noting, though, is that the hero in both is the fictional Marine Colonel Martin "Jiggs" Casey, who sees indications that some members of the Joint Chiefs of Staff are planning to take over the government and alerts the President to the plot. The President then thwarts the plot and forces the scheming Chiefs to resign. The military ethos (at least in the person of Colonel Casey) and civilian control ultimately prevail, even in fiction and film.

The superiority of the political over the military is not a notion unique to the United States or even to democratic societies. Carl von

Clausewitz, the professional military officer and theorist writing in early 19th century Prussia, argued that political considerations trump military "requirements":

> *Subordinating the political point of view to the military would be absurd, for it is policy that creates war. Policy is the guiding intelligence, and war only the instrument, not vice versa. No other possibility exists, then, than to subordinate the military point of view to the political.*[6]

For this to work today, military commanders, even at the most senior levels, must be subordinate to civilian political leaders, who formulate and implement policy in the name of the sovereign people. The Constitution, to which soldiers swear fealty, is in the end a compact of representative government.

To be sure, some countries today are run by their armed forces, but around the world the prevailing practice and, one might argue, the aspirational ideal is civilian control of the military, regardless of the nature of the political system. Notably, an important criterion for membership in the North Atlantic Treaty Organization is "the establishment of civilian and democratic control over military forces."[7]

While all U.S. military members, including officers, are sworn to support the Constitution's mandate of civilian control, the practical application of civilian control of the military plays out differentially within the officer corps. Junior officers have little *direct* engagement with civilian leaders; but as officers rise in rank, especially to general/flag officer levels and senior command and staff positions, interaction with civilian officials becomes a central part of their professional lives.

Civil-military interactions are influenced by cultural differences between the professional military and the civilian officials they serve. The distinguished military historian Russell Weigley traces the phenomenon back to colonial days:

> *From the beginning, career soldiers perceived themselves as occupying a somewhat hostile environment, distrusted by American civilians—which indeed they were, because American civilian culture had absorbed an English tradition inimical to standing*

> *armies even before any such armies appeared in the colonies that were to become the United States.*[8]

"The larger issue," Weigley continues, "is that historically American soldiers and civilians have always represented two different cultures."[9] Most U.S. Presidents and their senior political appointees have spent their adult lives as civilians, immersed in the civilian culture. So to some extent, the issue of civil-military relations, including civilian control of the military, falls in the realm of cultural anthropology, in that each side is to some extent a stranger to the other.

In the United States, the issue of *authority* is the easy part of civilian control of the military. Civilian superiority is enshrined in the Constitution and statutory law, and has been the prevailing practice for the life of the Constitution. More complex and challenging are issues of the relative influence and institutional power of civilian officials and of senior military officers. Weigley captures the essence of the problem:

> *The modern issue of civilian control . . . entails assuring* [sic] *that the military will not be able to use its bureaucratic influence and its claim to special expertise to bend larger national policy to the service of military institutional desires. . . . The danger to civil control was not anything so unsubtle as a coup, but rather that of a disproportionate military influence on policymaking, conditioned by an increasingly distinct (because professional) military interest.*[10]

The influence and power of the military institution, which is simultaneously both a profession, in terms of identity, and a bureaucratic organization, in form of structure, should not surprise any student of organizational behavior. As a profession, the military can overreach its legitimate area of special expertise, and as a bureaucracy, as any budget cycle demonstrates, military departments can distort national strategy through exercise of control over expenditure of significant resources. Max Weber identified the underlying phenomenon: "Under normal circumstances, the power position of a fully developed bureaucracy is always overtowering."[11]

Unequal Authority and Asymmetric Knowledge

One approach to framing discussion of the complexities and challenges of civilian control of the military is to think of it in terms of unequal authority and asymmetric expertise. As Richard Betts notes in his study of civil-military relations, "At issue is the tradeoff between control and expertise. Imbalance on either side may have positive or negative effects, depending on the particular values and expertise involved."[12] Eliot Cohen calls the resulting relationship "an unequal dialogue—a dialogue, in that both sides express their views bluntly, indeed, sometimes offensively, and not once but repeatedly—and unequal, in that the final authority of the civilian leader [is] unambiguous and unquestioned."[13] It is almost inevitable that tension arises between authority and expertise.

If the authority is unequal, its exercise is influenced by the practical requirement of each for the skills of the other. The expertise of civilian and military leaders is best described as "asymmetrical," meaning different in scope and content, rather than unequal.

Earlier portions of this book noted how the authority of the military professional rests upon the claim of extraordinary expertise in the application and management of large-scale deadly force, reflected in mastery of the technical capabilities (and limitations) of lethal and nonlethal weapons systems; in possession of a significant regional expertise and personal contacts; and in the unique ability to design and execute operational strategies and tactics deploying and employing military forces to achieve desired outcomes. At the same time, senior civilian leaders possess their own special knowledge and skills upon which the soldier depends for ultimate success. Generally speaking, senior civilian officials are likely to know more than senior military officers about such matters as the possibilities residing in international relations; economic-political connections; diplomatic arrangements and initiatives; U.S. and foreign domestic political considerations; and the array and manipulation of the capabilities of the various departments of the national government. These skills are critical to the development and execution of policy and strategy at the highest level. At the very top, they set the context and provide the rationale for the contribution of the military to national purposes. They give substance to the notion that *armed forces don't make war, nations do*. Most important, final responsibility for harnessing all the means of national power to achieve national ends resides with civilian officials, including the choice of ends and the decision to employ military forces.

Military officers should resist any temptation to insist that their opinions on such matters are superior to those of civilian political leaders. First, the subject matter most often exceeds the soldier's professional brief and competence, and second, as Samuel Huntington asserted, "No commonly accepted political values exist by which the military officer can prove to reasonable [people] that his political judgment is preferable to that of the statesmen."[14] Senior uniformed officers, with their distinct competitive advantages in military matters, must remind themselves that most crises and issues that rise to the highest levels, such as the National Security Council, are not uniquely and exclusively military in nature, and therefore that varieties of expertise in addition to—and not instead of—military expertise must be brought to bear in policymaking and decisionmaking. Multidimensional issues call for multidimensional solutions, which require meshing or integrating a rich variety of perspectives and skills within a particular policy perspective. Former Chairman of the Joint Chiefs of Staff General Richard B. Myers makes the point regarding the role the Secretary of Defense plays in reviewing operational plans:

> [One] *might think it was inappropriate for a civilian to say [he could] improve a military commander's plan. But the most critical elements in any operational plan were the assumptions that went into it. Many of these assumptions were political or geopolitical in nature, and therefore the Secretary would normally have great insight into their appropriateness.*[15]

Again, Clausewitz reminds us that "the nature of the political aim, the scale of demands put forward by either side, and the total political situation of one's own side, are all factors that in practice must decisively influence the conduct of war."

Clausewitz goes on to address what one might call "the division of labor" between professional military officers and their political masters:

> *We can now see that the assertion that a major military development, or the plan for one, should be a matter* for purely military

> opinion *is unacceptable and can be damaging. Nor indeed is it sensible to summon soldiers, as many governments do when they are planning a war, and ask them for* purely military advice. *But it makes even less sense for theoreticians to assert that all available military resources should be put at the disposal of the commander so that on their basis he can draw up purely military plans for a war or a campaign.*[16]

Making the Civil-Military Relationship Work

It is easier, of course, to describe this relationship of unequal authority and asymmetric expertise than it is to make it work effectively in the real world of policymaking and crisis management. In a thoughtful study of civil-military relations published in 2009, former Deputy Secretary of Defense John White and Sarah Sewall, who herself has held senior academic and government positions, captured the problem and pointed to ways to manage it more effectively:

> *In many respects, the civil-military relationship is an awkward construct. It demands the subordination of leaders in the military profession to civilians who, almost by definition, lack equivalent knowledge and expertise. It often forces civilians to make decisions on military issues by relying on their non-military knowledge even when analogies may not work; civilian leaders therefore require assistance from the military profession. The relationship requires that the two sets of actors divide their roles even as it becomes increasingly difficult in practice to differentiate between political and military judgments. It calls for partnership in the service of the Constitution even as individual actors face competing political, institutional, or Service loyalties.*[17]

Such a partnership must be built on mutual understanding, humility, and trust—characteristics intuitively admirable in principle, but which demand continual, difficult efforts by all parties, often in the most challenging circumstances where the stakes can be enormous, the costs and risks hard to specify, and the dangers formidable, sometimes imminent. Successful management of these challenges calls for *education* across that cultural and expertise gap: to help civilian officials understand practical military considerations (the "assistance from the military profession" that White and Sewall cite), and to help military leaders appreciate the broader context and complexities of the situation. Here the Nation's most senior military officers serve as the critical nexus. General Charles Boyd made the point sharply in an address to Air University, "Your task—indeed your responsibility—is to help them [civilian officials] make the right decisions. With all the power of persuasion you can muster, and at whatever personal risk you perceive that may require, you must tell your bosses what your professional judgment dictates."[18]

At the very top of the profession of arms, the U.S. Armed Forces are linked institutionally with the constitutional structure of government by the offices of the Chairman of the Joint Chiefs of Staff and the Joint Chiefs of Staff, under the authority of the Secretary of Defense and the President in his constitutional capacity of Commander in Chief of the Armed Forces. Though the Chairman and the Chiefs do not hold command, or exercise direct authority beyond their particular staffs, the Chairman is by law the senior uniformed officer of the U.S. Armed Forces, and the Chiefs of Staff are the senior officers of their respective Services. Under Title 10, the Chairman is "the principal military adviser to the President, the National Security Council, the Homeland Security Council, and the Secretary of Defense."[19] The Chairman and Service Chiefs of Staff provide the interface between professional competence and civilian authority, both in the structure of the separate military departments, and collectively as the Joint Chiefs of Staff.

As advisors, the Joint Chiefs are responsible for mediating the gap between the ambitions of policy and the limitations of military capability and, by the nature of their conduct, for guaranteeing the reliability of the members of the Armed Forces in adherence to their constitutional duty. As officials in an executive department of government, they are expected to support decisions with which they as advisors may have disagreed. All serve under the command of the President and Secretary of Defense; the Service chiefs serve under the authority of the Service secretaries who are the heads of the respective military departments. All are appointed by the President upon the recommendation of the Secretary of Defense and serve at the pleasure of the President. The President may dismiss any Chairman or any Service chief summarily.

At the same time, the appointments of the Chairman and the Joint Chiefs of Staff require confirmation by the Senate. The Chairman, who serves only a 2-year term, must be reconfirmed if nominated for a second. It has become a traditional part of the confirmation process for the Senate to require senior appointees to commit themselves in writing to offer their personal opinion to Congress, *if requested*, even if that opinion is contrary to the policy of the Commander in Chief, whose agents they are. In short, while the positions as military advisors to the President, the National Security Council, and the Secretary of Defense are established by Title 10, a corresponding responsibility to provide advice to Congress has grown up by convention, pursuant to the legislative branch's powers and authorities established in Article I of the U.S. Constitution.[20] Whether this involves a right to "lobby" Congress in opposition to the decisions of the executive branch remains an issue in practice, if not in theory. Notably, Franklin Roosevelt told Sam Rayburn, then the Speaker of the House, that a part of his respect for George Marshall, as Chief of Staff, derived from the fact that the President did not have to worry that Marshall would go to Congress to reverse the President's decisions. "I know he's going back to the War Department, to give me the most loyal support as chief of staff that any President could wish."[21]

The role of advisor is sufficiently vague to be the source of some controversy. Chairmen and Service chiefs are often excoriated for not speaking out publicly against government policies with which critics disagree, for technical, partisan, or ethical reasons. At other times they are blamed for not resigning in the glare of publicity, and for not then going to the country to oppose decisions of the Commander in Chief on grounds that appear compelling to particular critics, in and out of the Armed Forces. A proper antecedent question to judging these criticisms goes to the nature of professional advice within a system of *representative* government.

In January 2015, General Martin Dempsey, then Chairman of the Joint Chiefs of Staff, told Fox News Sunday News host Chris Wallace that his metrics for judging the relationship between elected leaders and their professional advisors are "access and whether my advice is—influences decisions."[22] Dempsey went on to indicate that he did not expect military advice to dictate presidential decisions, which are inevitably broader than military concerns, but said he knew he had access to the President and believed he could see that his advice did influence the President's subsequent actions.

A more thorough description of the Chairman's role was given 64 years earlier by General of the Army Omar Bradley, the first Chairman of the Joint Chiefs of Staff, in his 1951 testimony before the Senate committee inquiring into U.S. policy in the Korean War and the relief of General Douglas MacArthur. General Bradley addressed specifically the appropriateness of military professionals "speaking out" in opposition to government policy; the point where resignation by military advisors is appropriate; and the importance of confidentiality in communications with responsible civilian officials—three of the most common grounds for popular criticism of the Nation's military advisors.

Bradley addressed the limits of professional advice in response to a series of questions by Republican Senator Styles Bridges of New Hampshire. Bridges asked Bradley whose views should prevail in a disagreement about a military topic. Bradley countered that the particular issue mattered. Sometimes political and diplomatic issues legitimately had to prevail over military expediency. Bridges asked if, in that case, the military advisor ought not go to the public: "don't you think the American public are entitled to the best military judgment of our military leaders?" Bradley replied that the Chiefs' responsibility was limited to providing the best advice possible, and if it were not taken, there was nothing to be done. Bridges then asked: "If it reaches the time in this country where you think the political decision is affecting what you believe to be basically right militarily, what would you do?" To this Bradley said: "If after several instances in which the best military advice we could give was no longer of any help, why, I would quit. I feel that is the way you would have to do. Let them get some other military adviser whose advice apparently would be better or at least more acceptable."[23] Asked by Bridges if he would then speak out to the American people, Bradley replied he would not. "I am loyal to my country," he said, "but I am also loyal to the Constitution, and you have certain elected officials under the Constitution, and I wouldn't profess that my judgment was better than the President of the United States or the Administration."[24]

Bradley's testimony was subsequently interrupted when he refused to breach the confidentiality with which he advised the President. Bradley stated:

> *It seems to me, that in my position as an adviser, one of the military advisers to the President, and to anybody else in a position of authority who wants it, that if I have to publicize my recommendations and my discussions, that my value as an adviser is ruined . . . it seems to me that when any of us have to tell everything that we say in our position as an adviser, that we might just as well quit.*[25]

Bradley's assertion of confidentiality was ultimately acknowledged by the committee, after lengthy debate by committee members. It is important to observe that Bradley's objection here had to do with the *content* of advice offered the President and that he indicated a willingness to advise "anybody else in a position of authority who wants it."[26] Presumably Bradley's "anybody else in a position of authority" would include members of Congress exercising their responsibilities under Article I of the Constitution. It would not include anyone and everyone who asked for advice.

Bradley's principles are subject, like most constitutional questions, to various interpretations, as indicated by the committee debate on the limits of *confidentiality.*[27] The issue of what constitutes an appropriately "professional question" was not raised, though some of the most controversial issues involve precisely that question, especially those where professional judgment and/or constitutional authority are divided as, for example, in the life of the "Don't Ask, Don't Tell" policy of recent memory.[28]

There is also a question of the warrant of institutional jurisdiction over an issue inherently subjective among uniformed authorities. All military choices involve trade-offs, and exist in a realm of probability, not precision and certainty, and thus become questions of value as much as calculation. Any decision involves costs that may ultimately be grounds for criticism without, it seems, consideration of comparable benefits or even available alternatives. Aside from a range of civilian pundits, there is a large community of retired senior officers who claim continuing expertise without having any accompanying responsibility for confidentiality, objectivity, or results. The republic depends largely on the elected officials for choosing well where they seek their advice, whatever the provisions of law.

There is also the question of the extent of loyal obedience. It is a commonplace in the Armed Forces that free discussion is open and wide-ranging before the decision, but once the commander decides, the force "falls-in" and faithfully executes the decision. That practice, of course, is intended to end discussion that could detract from complete commitment to successful execution. In the case of the Chairman and Joint Chiefs of Staff, there is a kind of philosophic tension between their responsibilities to provide independent professional advice on military issues, and to serve as executive branch officials in the Department of Defense defending policy decisions with which they may have disagreed. Note that this is not a feature unique to the highest-level staff officer; the same tension applies to any officer at any level of command who has his or her recommendation overruled and then must defend and execute the commander's position with which the officer formerly disagreed.[29]

General Bradley indicated that professional opposition should end with registering disagreement with the appropriate constitutional authorities. The counter-case is that of Matthew Ridgway, who as Army Chief of Staff continued his public opposition to President Eisenhower's "New Look" military policy, which emphasized deterrence based on air-delivered nuclear weapons at the expense of the Army and Navy, even *after* the President had decided on the New Look policy. In this, Ridgway followed the precedent of the 1949 "Revolt of the Admirals," in which senior Navy officers were relieved by President Truman for opposition to a similar policy, though in this particular case, Eisenhower did not fire Ridgway. He simply did not reappoint the general, who had reached the age of retirement anyway, to a second 2-year term as Chief of Staff.[30]

Based on his experience, General Richard Myers draws critical lines:

> *In essence, the senior military officers' role is to vigorously provide the best professional military advice possible to our political leaders. The Commander in Chief or the Secretary of Defense*

makes the decisions. And unless they are illegal or immoral, the military must carry out the orders of the President or the Secretary. To do otherwise would be to impose our own military judgment on what are political decisions, an action that's fundamentally inconsistent with our Constitution or the laws of the land.[31]

In the end, the officer who cannot support the President's or Secretary of Defense's decisions in good conscience, or finds he or she has lost the ability to perform the advisory function of the office, must offer to resign, or as General Ron Fogleman chose, to retire. Writing in the late 1950s, Samuel Huntington addressed the most wrenching of cases, where the call of official duty and the call of conscience pull the officer in opposite directions:

For the officer this comes down to a choice between his own conscience on the one hand, and the good of the state, plus the professional virtue of obedience, upon the other. As a soldier, he owes obedience; as a man, he owes disobedience. Except in the most extreme instances it is reasonable to expect that he will adhere to the professional ethic and obey. Only rarely will the military man be justified in following the dictates of private conscience against the dual demand of military obedience and state welfare.[32]

Here Huntington reflects broader principles of public service ethics. Writing several decades after Huntington, Professor J. Patrick Dobel argues in *Public Integrity*[33] that the public official has to hold in balance three models, all of which have ethical wisdom and imperatives: the legal-institutional model, the personal responsibility model, and the effectiveness or implementation model. The first serves primarily to limit the discretion allowed public officials. The second serves to preclude any public official from saying "They made me do it."[34] The third points to the need for public servants to "achieve an excellent . . . outcome."[35] For the public servant, Dobel argues, the art is in balancing these three models, not in picking one over the others:

> *I believe that we should think about public discretion and integrity as an iterative process in which public officials move within a triangle of judgment. They move back and forth among the [three] domains . . . holding them in balanced tension when framing judgments.*[36]

Interestingly, Dobel's "balanced tension" echoes Clausewitz on his "remarkable trinity"—"primordial violence, hatred, and enmity," "the play of chance and creativity," and "subordination to policy." "Our task," states Clausewitz, "therefore is to develop a theory that maintains a balance between those three tendencies, like an object suspended between three magnets."[37]

By law, retired officers remain in the military establishment, on the retired list. If they choose to challenge the policy of the Commander in Chief publicly, they should consider the likely impact on the profession and the executive's confidence in those still in uniform, and consider with some humility that professional knowledge is subjective, transitory in detail, and highly contextual. In short, they should accept that their conclusions might be ill-informed and/or wrong. On the other hand, asked for their advice by responsible leaders, they, as much as General Bradley, are certainly bound to give it.

The essential element in making all this work, across the cultural and expertise gap, is *trust,* as historian Russell Weigley indicates: "Faithful military acceptance of civilian control is a major desideratum of the U.S. constitutional system. Better yet, however, is faithful obedience based on candid civil-military discussions and on mutual understanding and trust."[38] Nor is the point made only by academics. Thoughtful and successful practitioners are even more eloquent on the issue of relationships. John White and Sarah Sewall note that in their project, practitioners, civilian and military, "stressed the role of personal trust, and the need to constantly reinforce it given daily substantive and bureaucratic challenges to those relationships. . . . Trust was often described as the result of symbolic and concrete efforts they had personally made to demonstrate genuine interest in and respect for their partners in the relationship."[39] Experience shows that nurturing, even establishing, trust can be especially difficult in the early months of a new political administration, when civilian officials, unfamiliar with the professional military ethos, may question the loyalties of senior military officers who served under the previous administration. As the "junior partner" in this relationship, the burden is often on the military to make clear that their loyalties are dictated by their constitutional oath, and that they will faithfully serve whomever the American people choose as their President and Commander in Chief.

Notes

[1] *The Declaration of Independence and the Constitution of the United States of America* (Washington, DC: NDU Press, 1994), 32.

[2] Ibid., 54–56.

[3] Ibid., 66, 68.

[4] U.S. Code, Title 10—Armed Services, Section 162b, as quoted in Cornell University Law School, Legal Information Institute, available at <www.law.cornell.edu/uscode/text/10>.

[5] Morris Janowitz, *The Professional Soldier* (New York: The Free Press, 1960), 367.

[6] Carl Von Clausewitz, *On War*, ed. and trans. by Michael Howard and Peter Paret (Princeton: Princeton University Press, 1976), 607.

[7] *North Atlantic Treaty Organization Handbook* (Brussels, Belgium: NATO, 2006), 186.

[8] Russell F. Weigley, "The American Civil-Military Cultural Gap: A Historical Perspective, Colonial Times to the Present," *Soldiers and Civilians: The Civil-Military Gap and American National Security* (Cambridge, MA: MIT Press, 2001), 215.

[9] Ibid., 218.

[10] Ibid., 225.

[11] "Bureaucracy," in *From Max Weber: Essays in Sociology*, ed. H.H. Gerth and C. Wright Mills (New York: Oxford University Press, 1946), 232.

[12] Richard K. Betts, *Soldiers, Statesmen, and Cold War Crises* (New York: Columbia University Press, 1991), 41.

[13] Eliot A. Cohen, *Supreme Command* (New York: The Free Press, 2002), 209.

[14] Samuel P. Huntington, *The Soldier and the State: The Theory and Politics of Civil-Military Relations* (Cambridge, MA: Belknap Press, 1958), 76.

[15] Richard B. Myers, *Eyes on the Horizon* (New York: Threshold Editions, 2009), 135.

[16] Clausewitz, 607.

[17] John P. White and Sarah Sewall, *Parameters of Partnership: U.S. Civil-Military Relations in the 21st Century* (Cambridge, MA: The Harvard Kennedy School of Government, 2009), 10.

[18] "Air University Graduation, May 25, 2006: Remarks by General Charles G. Boyd, USAF (Ret.)," available at <www.au.af.mil/au/aul/img/boyd_speech_auGrad_2006.pdf>.

[19] Title 10, U.S. Code, Section 151, Joint Chiefs of Staff: Composition; Functions.

[20] Ibid.

[21] President Franklin D. Roosevelt, quoted in Charles F. Brower, *George C. Marshall: Servant of the American Nation*, ed. Charles F. Brower (New York: Palgrave Macmillan, 2011), 167.

[22] Martin E. Dempsey, interview with *Fox News Sunday*, January 11, 2015, available at <www.foxnews.com/transcript/2015/01/11/gen-dempsey-reacts-paris-attacks-sens-hoeven-coons-talk-keystone-showdown>.

[23] "Transcript of First Day of General Bradley's Testimony at Senate Foreign Policy Inquiry," *New York Times*, May 16, 1951, page 2 of 4 (photocopies), available online without pagination in the *New York Times* historical archive. Bradley's viewpoint was echoed in 1997 by General Ronald Fogleman who requested relief and expeditious retirement as Air Force Chief of Staff. Ronald R. Fogleman, "The Early Retirement of Gen Ronald R. Fogleman, Chief of Staff, United States Air Force," *Aerospace Power Journal* (Spring 2001), available at <www.airpower.maxwell.af.mil/airchronicles/apj/apj01/spr01/kohn.htm>.

[24] "Transcript of First Day of General Bradley's Testimony," page 2 of 4 (photocopies).

[25] Ibid., page 3 of 4.

[26] Ibid.

[27] "Excerpts from Proceedings in Joint Hearing on Dismissal of General MacArthur," *New York Times*, May 18, 1951, two pages (photocopies), available online without pagination in the *New York Times* historical archive.

[28] "A History of 'Don't Ask, Don't Tell,'" *Washington Post*, November 30, 2010, available at <www.washingtonpost.com/wp-srv/special/politics/dont-ask-dont-tell-timeline/>.

[29] See for example, General Bruce C. Clarke, "Integrity versus Professional Loyalty," *Military Review* 46, no. 8 (August 1966), 71.

[30] Andrew J. Bacevich, "The Paradox of Professionalism: Eisenhower, Ridgway, and the Challenge of Civilian Control, 1953-1955," *The Journal of Military History*, vol. 61 (April, 1997), 303-304; John S.D. Eisenhower, "How Generals Should Talk to Presidents," *New York Times*, October 17, 2009, available from *New York Times* electronic archive.

[31] Myers, 184.

[32] Huntington, 78.

[33] J. Patrick Dobel, *Public Integrity* (Baltimore: Johns Hopkins University Press, 1999).

[34] Ibid., 12.

[35] Ibid., 17.

[36] Ibid., 2.

[37] Clausewitz, 89.

[38] Weigley, 227.

[39] White and Sewall, 14-15.

CHAPTER EIGHT

The Officer and Society: The Horizontal Dimension

As chapter 7 explains, for the American Armed Forces officer, the vertical dimension of the profession of arms and society—civilian control of the military—is formally enshrined in the Constitution of the United States, which every officer is sworn to "support and defend." The drafters of the Constitution specified that all of the key powers regarding the military would be in the hands of civilian officials of the legislative, executive, and judicial branches of the Federal government. Over centuries of practice, civilian control of the military has been embedded in the American military's genetic makeup.

Equally important, but less well defined, is the *horizontal* dimension of the profession and society—how practices and values in the military Services mesh, or do not mesh, with practices and values in the larger society for whose "common defense" the Constitution was crafted and the Armed Forces created. Previous chapters have emphasized the importance of an officer's exemplary individual conduct to maintenance of effective civil-military harmony. This chapter focuses on the collective responsibility of the Armed Forces to keep their practices in harmony with the fundamental values of the parent society they serve.

A fundamental tension persists between the values that define a liberal democratic society such as the United States and the values that define the profession of arms. The former values seek to provide for the freedom and political equality of all citizens. The latter, in contrast, seek the effective *and disciplined* use of force in pursuit of national purposes. This requires subordination of the individual military member in ways that contrast significantly with the democratic doctrines of

American society. Generally speaking, contradictions or differences between the two diverging goals must be grounded in necessity and compatible with a broad understanding of and respect for the basic national values that the military Services are intended to secure.

The original expressions of American civic and political values are found in the Declaration of Independence and the Constitution of the United States. The Declaration asserts that:

- "all men are created equal, that they are endowed by their Creator with certain unalienable Rights, that among these are Life, Liberty, and the pursuit of Happiness"
- "Governments [derive] their just powers from the consent of the governed [and should] effect their Safety and Happiness."

While the Declaration articulates American ideals, the Constitution establishes the governing principles of the Nation, including civilian control of the military. It also spells out the Declaration's "unalienable Rights," most particularly in the Bill of Rights, which guarantees fundamental individual rights including freedom of religion, speech, press, and assembly; freedom to petition for the redress of grievances; the guarantee of a right to keep and bear arms; and the right of the people to be secure in their persons, houses, papers, and effects. More broadly, the Bill of Rights establishes provisions for protecting citizens from the powers of government. Although the Constitution prohibits the passage of any laws establishing religion, and guarantees all citizens its free exercise, it is otherwise an entirely secular document, which establishes an essentially secular government.

The profession of arms invokes and evokes other distinctive values, including those specified for the officer in the commission from the President of the United States: patriotism, valor, fidelity, and abilities, as well as strict performance of duty and obedience. Defined by their ultimate mission and purpose ("to provide for the common defense"), and by necessity hierarchical in nature, the U.S. Armed Forces call for certain sacrifices from their members, including giving up the free exercise of some of those rights and freedoms enshrined in the Constitution. When they put on the uniform, swear the oath, and accept a commission, officers voluntarily—and knowingly—accept limitations on their freedom of speech, limitations that would be anathema to their civilian fellow citizens. These rights may be "unalienable," but they can be forfeited or waived by the individual as a condition of service, and that is what officers do when they accept a commission. They end up in the paradoxical position of having sworn to defend their fellow citizens' constitutional rights, some of which they themselves have abjured for the common good.

Thus, some of the practices and values in the U.S. Armed Forces are noticeably and notably different from the practices and values of the larger, civilian society. Civilians, by and large, choose the cities, towns, and states where they want to live. Military members, in contrast, are issued *orders* that tell them where they will be living. Civilians regularly participate in public demonstrations for or against this, that, or another public policy, public official, or political candidate. In contrast, severe restrictions are in place on military members' freedom to wear the uniform in such demonstrations or speak in the person of their office in support of, or in opposition to, political questions of the day. Moreover, Article 88 of the Uniform Code of Military Justice forbids officers from expressing contempt for civil officials of both state and Federal governments. That these differences exist is true not only empirically, but also *normatively*. These differences are both necessary and desirable, as the U.S. Supreme Court noted:

> *This Court has long recognized that the military is, by necessity, a specialized society separate from civilian society. We have also recognized that the military has, again by necessity, developed laws and traditions of its own during its long history. The differences between the military and civilian communities result from the fact that "it is the primary business of armies and navies to fight or be ready to fight wars should the occasion arise."*[1]

At the same time, there are differences not only regarding practices, but also as to the importance and expectation of the presence of certain values and virtues. These differences were pointed out earlier in chapter 3 from the standpoint of the virtues inherent in military service. Here they are addressed again, in terms of the differences in their importance to civil and military societies.

The point is not that certain virtues abide *only* in military professionals. As General Sir John Hackett notes, "the military virtues are not in a class apart." He continues, quoting Arnold Toynbee, "they are virtues which are virtues in every walk of life . . . nonetheless virtues for being jewels set in blood and iron."[2] Physical courage offers one example of how such differences exist—and why they should. Courage is a noble virtue wherever and in whomever it appears. It is not, however, unique to soldiers. Acting bravely in the face of the enemy is admirable for civilians, but it is not expected from them, let alone mandatory. Civilians who fail to act courageously are not condemned. What is unique for the soldier, in contrast to the civilian, is not that bravery is esteemed, but that its opposite is condemned: cowardice in the face of the enemy is punishable by court martial, and is perhaps the military equivalent of a mortal sin.

What is true for physical courage is true for many other virtues as well, virtues that are integral to the profession of arms. To quote Hackett again:

> *What is important about such qualities as these ... is that they acquire in the military context, in addition to their moral significance, a functional significance as well. . . . Thus while you may indeed hope to meet these virtues in every walk of life, . . . in the profession of arms they are functionally indispensible* [sic].[3]

"Soldiers need virtues," asserts David Fisher after citing Hackett, "to make them effective soldiers."[4] What the civilian ideally should be, military officers *must be*, if they are to fulfill the obligations of subordination and service to which they are committed.

At the same time, if values and practices in the military, and those in the larger society, either drift or march too far apart, then the living tissue that binds the two together is stretched or even torn, with adverse consequences for both. The extent and severity of such differences, and how to reconcile them, have been discussed and debated throughout the Nation's history. The issue remains less than completely and definitively resolved, probably because no absolute, permanent resolution is possible. How to reconcile those differences, and how best to balance the two sets of values, is a perennial, political, and practical challenge for the military, especially its officers, and for the society, especially its civilian leadership. Some underlying harmony between the Armed

Forces and society is not only desired, but also necessary for the effective defense of the Nation, the existential purpose of both the Armed Forces, and largely, the Federal government.

Their experience with the British "Redcoats" stood out in the minds of the signers of the Declaration of Independence, and it was not salutary. Indeed, it was one of chasms that lay wide and deep between the people of the colonies and the government against which they were rebelling:

- "He has kept among us, in times of peace, Standing Armies without the Consent of our legislatures."
- "He has affected to render the Military independent of and superior to the Civil Power."
- He has "quartered large bodies of armed troops among us."[5]

This was a reality the Founders strove consciously to avoid as they established their new government; indeed, they wanted something quite different. Thus, deeply rooted in the American DNA is the belief that the American Armed Forces should *come from* and *be anchored in*, and *not alien to* the American people. "The relationship of the Armed Forces with the American people is both pragmatic and moral."[6] The Armed Forces rely on the American people to set the conditions under which men and women, America's sons and daughters, join and serve in the military Services and "wear the cloth of the nation"; to fund military salaries, and the equipment, training, health care, and housing military personnel require; to support them from afar when they are sent into harm's way; and to provide for their long-term care through the Department of Veterans Affairs. Without the active, continuing, tangible support of the American people, the Armed Forces would wither and disappear, no longer able to "provide for the common defense."

The moral connection is more critical than the pragmatic. This is the sacred bond of trust, the trust that gives the American people confidence that the members of the Armed Forces will "provide for the common defense" through reliable, competent, effective, efficient performance of their duties and, reciprocally, gives the men and women in uniform confidence that the American people respect their service and the sacrifices they make, and will "have their backs" in war as well as peace. These are the proverbial ties that bind the American people and those who serve them in uniform—their Soldiers, Marines, Sailors, Airmen, and Coastguardsmen.

The compact between the people of any nation and the professions that serve them is built and nurtured on mutual recognition of shared values and acknowledgment of natural, necessary differences. Managing the balance between the two is an art, not a science. These shared values come from the Nation and its people, and the professions and their members must adopt those values, internalize them, and incorporate them into their own professional values—if they are to maintain the trust of those they serve. As Brigadier General Anthony E. Hartle has described the relationship, "the subset of national values that we must identify are moral values, those that have an interpersonal focus or that concern good and bad character. The moral values of society will exercise the major influence on the content of particular ethical codes within that society."[7] In the case of the United States, Hartle said, those "moral values of society" include democracy, freedom, individual integrity and dignity, and equality in terms of rights. The American Armed Forces are sworn to protect those values, and in order to maintain the trust of the American people, they must embody them to the greatest extent consistent with their professional obligations as members of the profession of arms.

The ideal relationship between a profession and the society it serves is one of "moral integration." As James Burk argues, citing the work of Edward Shils:[8]

> *societies have a central value system that informs expectations about how institutions should conduct themselves if they are acting properly or legitimately. When institutions conduct their business and maintain relations with society that accord with those expectations, then we can say that the institution is morally integrated with society.*[9]

"Moral integration with society," Burk continues, "is a key element of organizational legitimacy."[10] Legitimacy, in turn, is a key element in building and nurturing trust between an institution or profession and the society and the people it serves.

Military-Civilian Gap

If practices and values in the military and those in society become less harmonious and if they drift or march too far apart, then the desired and vital moral integration deteriorates. At the end of the Cold War, both American society and the Armed Forces struggled to redefine their place in the world. As the Soviet Union, the principal antagonist against which most of the American military had prepared for 45 years, disintegrated and withdrew into vastly reduced borders and circumstances, a certain amount of introspection and reflection developed in the American defense community. One source of external concern had to do with academic criticism of what appeared to be the exercise of undue professional involvement in foreign and defense policy issues, accompanied by a growing tendency toward public expression of partisan preferences by members of the Armed Forces, and by the very public participation of retired officers offering endorsements in partisan political conventions and public criticism of defense and foreign policies in the 24-hour broadcast media—phenomena that continue today. On the other side, there was concern within the political leadership and the uniformed military that the mutual understanding essential to effective moral integration of the Armed Forces and general public was beginning to fray, due to a lack of familiarity on both sides and a belief on the public side that the requirement for military forces had disappeared and a "peace dividend" was to be expected.

Senior U.S. officials aired concerns publicly, lamenting the loss of civil-military moral integration. William Cohen, then Secretary of Defense, raised this worrisome prospect in a September 1997 speech at Yale University:

> *So one of the challenges for me is to somehow prevent a chasm from developing between the military and civilian worlds, where the civilian world doesn't [fully] grasp the mission of the military, and the military doesn't understand why the memories of our citizens and civilian policy makers are so short, or why the criticism is so quick and so unrelenting.*[11]

Just 2 years later, Richard Danzig, who had served as Secretary of the Navy, noted the damage that would ensue if the military and society lose their moral integration: "To allow the military services to drift away from the society that must nurture them is to put great institutions in great jeopardy."[12] In 2011, Representative Ike Skelton, long-time member and former chairman of the House Armed Services Committee, told an audience of one-star military officers from all five Services: "First, there is a military-civilian gap, it is serious, and it is growing. Second, there are two sides to this gap. Both the military and society have contributed to the creation and expansion of this gap. Consequently, there is work that must be done on both sides in an effort to narrow this gap."[13]

These worries were not confined to civilian officials. Senior military officers expressed similar concerns. In an address to a January 2011 conference on military professionalism held at the National Defense University, Admiral Mike Mullen, then Chairman of the Joint Chiefs of Staff, articulated the same concern:

> *But our audience, our underpinning, our authorities—everything we are, everything we do comes from the American people. . . . And we cannot afford to be out of touch with them. And to the degree we are out of touch, I think it's a very dangerous course . . . we don't know the American people. The American people don't know us. And we cannot survive without their support—across the board.*[14]

While harmony and moral integration between the two cultures are the ideal, the question remains: how can society manage those areas in which military and societal practices and values differ? One view is that the integrity of the military as a profession, and the value of preserving its ethos intact, argue, indeed even demand, that society not only not tolerate such differences as exist but, that in order to maintain a desirable degree of harmony, society must adapt its practices and values to correspond more closely to those of the military. This was the position taken in the depths of the Cold War (1957) by Samuel Huntington in *The Soldier and the State*. After laying out some of the differences between the profession of arms and a liberal democracy, he argued: "The requisite for military security is a shift in basic American values from [classical] liberalism to [classical] conservatism. . . . If the civilians permit the soldiers to adhere to the military standard, the nations themselves may eventually find redemption and security in making that standard their own."[15] In short, Huntington argued that all would be well if civilians would only act more like the military.

Taking a quite different, perhaps somewhat more nuanced and less "pure" position was Huntington's contemporary, Morris Janowitz. He noted "a convergence of military and civilian organization: the interpenetration of the civilian and the military is required. . . . It has become appropriate to speak of the 'civilianization' of the military profession and of the parallel penetration of military forms into civilian social structures."[16] In the original (1960) edition of his book, *The Professional Soldier*, Janowitz argued that even traditional military virtues have had to adapt to societal norms, that is, norms from *outside* the profession: "Military honor has had to respond . . . to changes in the social values in the society at large."[17]

History reveals that the Huntington view has not prevailed. Indeed, in a democratic society grounded on individual liberty, it was unlikely to do so. What has happened, over time, looks more like Janowitz's notion of convergence. Three descriptive models have emerged that explain how changes in values and practices in the military have occurred in relation to changes in values and practices in the civilian society since World War II.

Models of Military-Civilian Integration

In the first model, practices in the military were forced to change well in advance of changes occurring in the larger society. This was the case with regard to racial integration of the Armed Forces. On July 26, 1948, President Harry Truman issued an executive order intended to end segregation by race in U.S. military units:

> *It is hereby declared to be the policy of the President that there shall be equality of treatment and opportunity for all persons in the armed services without regard to race, color, religion or national origin. This policy shall be put into effect as rapidly as possible, having due regard to the time required to effectuate any necessary changes without impairing efficiency or morale.*[18]

Truman's order came 6 years before the Supreme Court declared in *Brown vs. Board of Education* that racial segregation in public schools was unconstitutional, and therefore must end. Racial integration in the military remained uneven until the Korean War, when military necessity forced changes in deeply ingrained, decades-old practices that mirrored practices of racial segregation in the larger American society.

Racial integration in the military was not seamless or trouble-free. Thoughtful, determined leadership—at all levels in the chain of command—was required to facilitate that monumental transition. Likewise in the broader society, successive Supreme Court decisions, sometimes enforced by Federal military and police powers, were required to bring about the end of racial segregation in public schools, after desegregation by "all deliberate speed" proved to be neither deliberate nor speedy. And it took a stormy, but inspiring, decade of civil rights activity and legislation to implement racial integration across all domains of American life. Here too, the process was not seamless or trouble-free, but the trajectory was clear. In the military Services, racial integration is largely a success story. In civil society, the struggle still continues.

In a second model of military-civilian integration, practices in the military changed in parallel with changes in practices in the larger civilian world. This was the case of expanding gender opportunities in the 1970s. During that decade, Congress dispensed with separate organizational structures for women and mandated that opportunities in the military previously denied to women must now be made available to them, perhaps most notably allowing women to enter the U.S. Service academies. At that time exceptions were made for those specialties involving direct combat. Career paths opened for women in uniform as opportunities for women were expanding outside the Armed Forces in higher education, athletics, and the corporate world. As in the case of racial integration, practical factors played a part in facilitating this transition. With the end of conscription and the introduction of the All-Volunteer Force, there was widespread concern within the Services that they would be unable to enlist the necessary number of recruits *of sufficient quality* to "man" and command the force. That meant that the Services needed to draw from a wider pool of candidates—women as well as men.

Once again, this sea-change transition was not always smooth and easy, but the overall vector was clear and enormous progress was made. Recent orders by the Secretary of Defense to open *all* service specialties, including combat arms, to qualified women indicate that the process of gender integration is still ongoing and contested.[19] Most often the remaining issue is the identification and achievement of consensus on credible standards to define who is *qualified* for particular roles. Notably, the profession has a role in advising the civilian authorities on these matters, but the final decision rests with the civilian masters who retain the constitutional authority to tell the professionals who will be allowed to serve.

More recently, a third model has appeared—where changes in the military lag behind changes in major segments of the broader society. This is the case of discrimination based on sexual orientation. In the second decade of the 21st century, following a huge shift in civilian attitudes, Congress repealed a 20-year-old statute that enabled the Defense Department policy of "Don't ask/Don't tell," a policy which permitted service by gay, bisexual, and lesbian military members only so long as their sexual orientation did not become known. Repeal of the statute led to a policy decision by the executive branch to allow gay, bisexual, and lesbian Servicemembers to serve openly in the Armed Forces of the United States, a transition achieved with remarkable speed by all the Services.

The question of the status of transgender Servicemembers remains in contention in civil society. For the military Services, the question has been answered by the civilian authorities at the top of the Department of Defense.[20] Subject to unexpected challenges from the Congress under its Article I powers, this policy seems unlikely to be reversed. As in the other cases mentioned, it now becomes the duty of the officer corps to provide leadership and wisdom to produce an effective armed force from all those persons the civil government deems eligible for service. At this writing, the military departments are engaged in doing so.

As shown above, change in all three of these models was neither seamless nor trouble-free. All three posed and continue to pose leadership challenges at all levels of the chain of command as deeply held individual values clash with wider public and Service values; none of those challenges, though, need be insurmountable. *Leadership matters.* Even though problems exist, and will likely continue, the trajectory of change is clear: practices (and ultimately the values they reflect) in the military and in the parent society must achieve more, rather than less, *moral integration.*

Officers bear special responsibilities to ensure that duly enacted laws and properly established policies are enforced, internalized, and followed in the day-to-day lives of the men and women in uniform. Those finally unable or unwilling to adapt must be identified and separated from the Armed Forces for the health and integrity of the profession. This is often not easy, and sometimes quite difficult, but officers must ensure that the laws and policies they are sworn and commissioned to uphold are implemented properly, even officers who might have personal, private objections to some of those laws and policies. Commissioned officers, military men and women serving under authority, do not get to choose which laws and which policies they will carry out. The moral obligations of the oath and commission must be respected: the officer must do his or her duty in spite of personal belief, or take leave of the profession. There is no third way.

Balancing the requirements and imperatives of the profession of arms and the values and ideals of a liberal democratic society like the United States is an art, not a science, and calls for continual monitoring, attention, and leadership. The stakes are enormous for both the military and the society it serves: maintaining and nurturing that bond of trust the Founders insisted upon for the new, very different nation and armed forces they were building—a bond of trust that is the polar opposite of the relationship between the Redcoats and the colonists in the 18th century. That delicate, challenging work continues, and Armed Forces officers, with and because of the "special trust and confidence" placed in them by the President of the United States, must be in the forefront of those efforts.

Notes

[1] United States ex rel. Toth v. Quarles, 350 U.S. 11, 350 U.S. 17 (1955), available at <https://supreme.justia.com/cases/federal/us/417/733/case.html>.

[2] Sir John Winthrop Hackett, *The Profession of Arms* (New York: Macmillan, 1983), 141.

[3] Ibid., 141.

[4] David Fisher, *Morality and War* (Oxford, UK: Oxford University Press, 2001), 109.

[5] *The Declaration of Independence and the Constitution of the United States of America* (Washington, DC: NDU Press, 1994), 32–33.

[6] Richard Swain, *The Obligations of Military Professionalism*, paper commissioned by the Institute for National Security Ethics and Leadership, National Defense University, December 2010, 7.

[7] Anthony E. Hartle, *Moral Issues in Military Decision Making*, 2nd ed., rev. (Lawrence: University Press of Kansas, 2004), 136.

[8] Edward A. Shils, *Center and Periphery* (Chicago: University of Chicago Press, 1975).

[9] James Burk, "The Military's Presence in American Society, 1950-2000," in *Soldiers and Civilians: The Civil-Military Gap and American National Security*, ed. Peter D. Feaver and Richard H. Kohn (Cambridge, MA: MIT Press, 2001), 250.

[10] Ibid., 262.

[11] William S. Cohen, "Remarks as delivered by Secretary of Defense William S. Cohen Yale University, New Haven, Connecticut, September 26, 1997," News Release, Office of Assistant Secretary of Defense (Public Affairs), No. 562-97, October 22, 1997.

[12] Richard Danzig, *The Big Three: Our Greatest Security Risks and How to Address Them* (Washington, DC: NDU Press, 1999), 55–56.

[13] Ike Skelton, unpublished remarks prepared for delivery at graduation dinner for Fellows of the National Defense University's Capstone course.

[14] Chairman of the Joint Chiefs of Staff, speech at the National Defense University Conference on Military Professionalism, as delivered by Admiral Mike Mullen, January 10, 2011, available at <www.jcs.mil/speech.aspx?ID=1517>.

[15] Samuel P. Huntington, *The Soldier and the State: The Theory and Politics of Civil-Military Relations* (Cambridge, MA: Belknap Press, 1957), 464, 466.

[16] Morris Janowitz, *The Professional Soldier* (New York: The Free Press, 1971), xi.

[17] Ibid., 217.

[18] Executive Order 9981, "Establishing the President's Committee on Equality of Treatment and Opportunity in the Armed Forces," available at <www.trumanlibrary.org/9981a.htm>.

[19] Secretary of Defense, Memorandum for Secretaries of the Military Departments, Acting Under Secretary of Defense for Personnel and Readiness, Chiefs of the Military Services, and Commander, U.S. Special Operations Command, Subject: Implementation Guidance for the Full Integration of Women in the Armed Forces, December 3, 2015, available at <www.defense.gov/Portals/1/Documents/pubs/OSD014303-15.pdf>.

[20] Memorandum for Secretaries of the Military Departments, Directive-type Memorandum 16-005, "Military Service of Transgender Service Members," Department of Defense, June 30, 2016, available at <www.dtic.mil/whs/directives/corres/pdf/DTM-16-005.pdf>.

CHAPTER NINE

Service Identity and Joint Warfighting

The Armed Forces of the United States consist of five military Services—the Army, Marine Corps, Navy, Air Force, and Coast Guard. In the 21st century, the days of any Service operating as a truly independent actor are long since past. The five Services fight together as a team, which means they must plan and train as a team. That does not mean that all five play equal parts in every battle or exercise. It does mean that the five are partners in the overall business of defending the United States, its territory, population, and national interests, and, therefore, that the best each Service has to offer must be woven into every battle, exercise, and plan. There can be no "lone wolves" among the five Services, because our security cannot afford free agents. When the Nation is threatened, the *Navy* doesn't go to war, nor does the *Army*; the *Nation* goes to war, using all its Services' capabilities in the combination that best suits the particular threat posed and the war plan designed to defeat it.

While "jointness" has become the short-hand description for this five-Service partnership (with its own "color"—purple), there is another way to characterize the relationship among the Services, one with deep roots in American history and political culture: *E pluribus unum*—From many, one. Inscribed on the banner held in the beak of the eagle on the Great Seal of the United States, approved by Congress on June 20, 1782, those words convey the reality that out of the original 13 colonies, one Nation emerged. The 13 new states kept their own identities, as well as their own local customs, food preferences, accents, and so forth, but together they constituted one Nation that was not just the sum of the 13, but greater than the combined total.

So, too, from five Services comes the one entity—the Armed Forces of the United States—charged with the defense of the Nation.

Tradition and identity, including uniforms and customs, matter, as do the requirements generated by the distinctive roles the various Services perform; the requirements involved in operating and fighting on land, at sea, in the air, in space, and in the cyber realm; and the different capabilities they bring to the battle. Thus, the Services keep their separate traditions and identities, their distinctive uniforms and customs; but out of the five of them emerges a single armed force that, because of the synergies among them, is greater, more flexible, and more capable than the mere sum of the five.

This book is all about being an officer in the Armed Forces of the United States in the 21st century. That involves being an officer in the Army, Marine Corps, Navy, Air Force, or Coast Guard, while also being an officer in something larger—the Armed Forces of the United States. Being a fully effective officer, both in one Service and in the Armed Forces, requires knowing one's own Service well, including its capabilities and limitations, and knowing the other Services well enough to appreciate their strengths and weaknesses, what they bring to the fight, and how their capabilities can best mesh with those of the other Services.

Each Service has its own uniforms, customs, and traditions. On a deeper level, each has its own culture. It is *culture* that defines and describes any organization best. It also best defines and describes what it means to be a member of that organization. As used here, culture is taken to have two meanings: on the organizational level, how this Service defines and sees itself; and, on the individual level, what it means to be a Soldier, Marine, Sailor, Airman, or Coastguardsman.

Thus, this chapter's contribution to understanding what it means to be an officer in the Armed Forces is to capture, albeit in snapshot style, what it means to be an officer in each of the five Services. To that end, we present five short essays written by former Service chiefs, each of whom describes what it means to be an officer in that Service.

The U.S. Army, by General George W. Casey, Jr., USA (Ret.), Chief of Staff, 2007–2011

To be an Army officer means that you are a leader. You bear the sacred responsibility of leading men and women in the most demanding of all human endeavors—ground combat. Their very lives depend on you. So it's no wonder that we invest so heavily in the development of our officers, warrant officers, noncommissioned officers, and civilians.

Three traits—*vision*, *courage*, and *character*—will form the essence of effective military leaders in the years ahead.

Vision. The primary function of any leader is to point the way ahead. This requires the ability to "see around corners"—to see something significant about the future that isn't readily apparent to others. The more volatile and the more ambiguous the environment, the harder it is for leaders to come to grips with the situation themselves—let alone articulate a clear way ahead. Today's volatile environments become invitations for inaction—people become befuddled by the complexity and uncertainty and don't act. So in today's environments, it is even more important for leaders to provide a clear vision to drive their organizations' actions.

The Number One question any commander should ask is: "What are we really trying to accomplish?" The higher in the organization one is, the more complex issues become, and the harder it is to answer that question clearly and succinctly. Senior leaders must get clarity in their own minds so they can clearly articulate to subordinates how they see things and what they want their subordinates to do. Nothing gets clearer when it leaves higher headquarters and begins trickling down through the many layers of the chain of command. If it isn't clear coming from the top, the poor Soldier on the ground doesn't have a chance.

Effective action begins with a clear statement of what needs to be accomplished. The clearer the commander can be, the better the subordinates will execute—even if this concept is not exactly right. Without a clear focus, there is no common purpose, and without common purpose, there isn't effective execution. In war that is fatal. The clearer leaders can be about what they want to accomplish, the better their organizations will execute in the volatility, uncertainty, complexity, and ambiguity of today's global environment.

Courage. Developing and articulating a clear view of the future in today's increasingly complex environments is hard work for a lot of reasons, but most importantly because it demands that leaders make judgments about the future— something that, because we are human, always entails risk. We could be wrong, and there could be significant consequences.

That's why it takes courage to lead—and it always has. Nothing good happens without risk, and it takes courage to act in the face of uncertainty and risk. And to succeed you must act.

Leaders must think things through, speak their minds, take action, and if they make a mistake—which we all will—must be resilient enough to adapt and bounce back. Acting is more important than not being wrong.

Character. Our 26th President, Theodore Roosevelt, stated, "Alike for the nation and the individual, the one indispensable requisite is character." He was talking about the positive mental and moral qualities distinct to an individual.

Leaders with strong values build strong organizations. In the Army, because the stakes are so high, we place significant emphasis on building character. From the day Soldiers enter Basic Training, the Seven Army Values are instilled in them—Loyalty, Duty, Respect, Selfless Service, Honor, Integrity, and Personal Courage. These values form the basis for morally strong and ethical Soldiers and leaders. They form the core of our leaders' character.

Character is most important in the leader. People trust men and women of character because they know that they will do the right thing for the organization and not themselves when the going gets tough; and that trust becomes the glue that binds organizations together.

Writing around 340 BC, Aristotle stated that moral goodness (character) is the result of habit. If you do good things repeatedly, you will be a good person. I found that to be true over my 41-year career. As I made decisions as a young officer on simple (in retrospect) matters, I developed the habit of doing what I thought was best for the organization I was serving and acting with conviction—something that prepared me for the very difficult choices I had to make as the commander in Iraq and Army Chief of Staff. Good character is essential, and building it starts early. Acting with conviction builds credibility.

Being a leader is the essence of being an Army officer. Vision, courage, and character are the traits that will make Army officers successful in the 21st century.

The U.S. Marine Corps, by General James T. Conway, USMC (Ret.), Commandant, 2006–2010

During World War I, General John J. Pershing recognized the superior bearing and discipline of Marine units, asking frequently, "Why in Hell cant [sic] the Army do it if the Marines can; they are all the same kind of men, why cant [sic] they be like Marines?"[1] The general's question was entirely logical in 1918—and seemingly would be as valid today. Marines and other U.S. Servicemembers come from the same towns, suburbs, and cities. They are all products of the same American culture and generally share the same values. Their officers graduate from the same campuses and swear identical oaths to the Constitution—so how divergent can they be? Let there be no doubt, Marines and their officers *are* different.

It starts with the Corps' culture, which has immediate and lasting impact on every man and woman who has earned the right to wear the Eagle, Globe, and Anchor. From the earliest days at Parris Island, San Diego, or the hills of Quantico, Marines are taught instantaneous obedience to orders, the importance of the mission as the highest priority, and the value of the team over the individual. They are taught that of all character traits, *integrity* is by far the most important—both on and off the battlefield.

The training is also different and is incredibly physical. Legendary football coach Vince Lombardi's admonition to his Green Bay Packers team—"Fatigue makes cowards of us all!"—is growled daily by drill instructors as if to ward off evil spirits in hardening bodies. Weapons proficiency and marksmanship build esteem and confidence. Not least, Marines absorb the rich history and traditions of the Corps, and are taught that perhaps the greatest sin in life would be to somehow tarnish the legacy of those Marines who have gone before.

For officers the training is, if anything, even more physical. It has to be, because the role of the officer in the Fleet Marine Forces is to think ahead when everyone else is sucking wind. In some professions the colloquialism is "never let them see you sweat." For the officer of Marines, it is "*always* let them see you sweat."

Officers learn the hard skills of their trade, the sophisticated nuances of their chosen military occupational specialty (MOS), and how to apply "tough love" to their Marines. They quickly see that in the best units officers and enlisted push each other, with each reveling in the success of the other.

The average age of Marines is dramatically younger than in the other Services. The mission requires it. But the difference also lends itself to the concept of father-to-son/teacher-to-student relationships. Marine officers welcome the respect that comes with their rank and responsibility, but work to ensure that respect is returned in full to those down the chain of command—from staff noncommissioned officers (NCOs), who are the backbone of any organization, to NCOs who will lead small units into battle, and finally to troops who will do the "heavy lifting."

Throughout their time in uniform, be it 3 years or 30, Marines relish the challenges associated with "doing more with less," and those challenges are many. Budgets are invariably tight, and conditions on bases and stations are Spartan in comparison to our brother Services. But the analogy does not end there. Marines realize that when war comes, the role of the Corps will be like that of the Spartans in the Greek phalanx—at the point of the spear or facing the most capable enemy. *Esprit de Corps* and a sense of elitism grow well in such an environment and breed conviction amongst all ranks that their word is their bond, that a handshake with a fellow Marine is more binding than a signature, and that if the worst happens, "I've got your back."

In the final analysis, Marine officers are different from their peers in the other Services for four reasons. First, every Marine officer during entry-level training—whether a pilot, logistician, or lawyer—is drilled for 6 months on how to command a Marine rifle platoon. Those basic combat skills are particularly invaluable today when the enemy can be anywhere and linear battle lines no longer exist.

Second, every officer, regardless of MOS, has a singleness of purpose: To enable, support, or lead grim-faced 19-year-old Lance Corporals so that they can destroy the Nation's enemies. Everything else in the Corps is secondary to that primary function.

Third, a Marine officer has a tremendous sense of the Corps' history and therefore a personal responsibility to maintain the legacy of his or her unit. The officer is driven by a resolute belief that if there is ever to be calamity, it will not happen on his or her watch!

Fourth, Marine officers have every confidence that they will be deployed into some God-forsaken patch of earth where they and the unit will have to adapt and overcome both the environment and the enemy. It is the nature of an amphibious or expeditionary force—and fortune favors those who have prepared themselves, mentally and physically, for uncertainty.

And yes, there is perhaps a fifth difference: All Marine officers, on active duty and for the rest of their lives, treasure that they have had the opportunity to lead some of the finest young men and women America has ever produced—those who wear the uniform of United States Marines.

The U.S. Navy, by Admiral Gary Roughead, USN (Ret.), Chief of Naval Operations, 2007–2011

The history of the military Services and their inspiring narratives are written in epic battles and accounts of extraordinary heroism and great sacrifice in time of war. The U.S. Navy is no different. But the history of the Navy and its foundations are found both in war, where it has fought and won our Nation's wars, and in peace, where it continues to ensure the international flow of resources and goods that determine the prosperity of our country. Whether in peace or war, the U.S. Navy is the Service that is always forward on the sea-lanes of the world, in places where our national interests and those of our allies and like-minded countries exist. It is the enduring, persistent, and consequential global mission of the Navy that underpins its structure, character, and tone.

Navies are capital-intensive: the ships, submarines, airplanes, and the infrastructure that supports them represent significant investments. Accordingly, the power, range, and capacity of the Navy are a function of what the Nation buys. What and how much are determined by our national interests, global obligations, assessments of geopolitical and technical trends, the Nation's technical and industrial capacity, and the political determination of how much to invest in the Navy.

Those expensive things have value only when the human element is added—the men and women of the U.S. Navy. They define how and how well the Navy performs as a function of culture, ethos, tradition, and competence. Those aspects are shaped over time by experience, standards, and the norms and expectations of the society from which we draw those who serve. This is a complex mix to be sure, but there are some factors that stand out that have shaped and continue to define the Navy.

Global Maneuver, Global Response. Nothing influences an organization over time more than the environment in which it exists and operates. For a military service, that environment shapes its organization, practices, traditions, and character. The Navy's operational environment is the vast oceans of the world—70 percent of the earth's surface and an international commons. Moving naval power without having to seek and obtain basing or overflight permission, especially when concerns of sovereignty are more acute in today's connected world, continues a tradition and *duty* of being prepared to be first on the scene in the Nation's response to crisis and conflict. To remain poised persistently in international space, unencumbered by political sensitivities of being on the ground, reinforces the imperative for operational flexibility that other Services do not enjoy. That unencumbered maneuver space of the sea is also changing and unforgiving in its natural power.

These are the factors that define and shape navies and the officers who serve in them. There is an allure to such an operational and physical environment, but there must exist an aptitude and a personal and professional comfort in operating, fighting, and leading at sea. Accordingly, a naval officer must be the following:

Uniquely Independent and Self-Reliant. Times and technology have changed and the complete (and enviable) autonomy enjoyed by those who put to sea in pre-wireless days is gone forever. However, regardless of how well connected we may be today, a small ship on a large ocean reinforces the Navy's culture of independence and self-reliance. Implicit in this is the concept of *shipmate*, every man and woman onboard depending on and trusting in each other for victory, success, and safety. Nevertheless, that same environment reinforces the importance of a culture of command in the Navy and the imperative of accountability.

Confident in a Culture of Command. Environment, independence, and self-reliance are the basis for the Navy's culture of command and accountability. This approach is unique to those who go down to the sea in ships and predates the founding of our Navy. The linkage of responsibility, authority, and accountability that endures today was forged in the unforgiving environment of the sea, the self-contained community of the ship, and the need for life-and-death decisions at any time. Thus exists an overriding priority on command, not a staff approach to decisions, and with it a seemingly obsessive sense of accountability. To many it may seem unjust and too unforgiving, but that is the essence of the culture of command and the "cruel business of accountability"[2] at sea and within the Navy.

At Ease in Life in Five Dimensions. The complexity of warfare continues to increase. Precision, lethality, range, and domains expand. In the last century navies have evolved from fighting in the single dimension of the ocean's surface to three dimensions—on, above, and below the surface. Add to that now space and cyberspace. A single ship can reach into each of the domains and be threatened from any of them. Technology, like the tide, will not be turned back, and innovation and experimentation will be the difference in whether and how the Navy will win or lose. The proud communities within the Navy retain their identities, but the need for integrated, whole, and innovative solutions will prevail.

Defined by Deployments. All Services have been stressed recently by repeated deployments associated with our recent wars, and the burden and costs have been heavy on all. But the Navy is a Service that deploys routinely. That is what it does and has done for centuries. The Navy is about being forward in war *and* peace. Only forward presence enables the swift, decisive response the Nation has come to expect of its Navy in crisis, conflict, or disaster. Importantly, the Navy's culture of deployment is an obligation shared unstintingly by those who remain behind—Navy families.

Dauntless in Diplomacy. A U.S. Navy warship is the United States. On the high seas or in a foreign port, that ship is the United States of America—our flag is flying there; it is U.S. sovereign territory. Whether commanded by a lieutenant or a captain, that ship is a statement of our presence, interests, and resolve. Every Sailor represents the Navy and the Nation, and he or she knows it.

Committed to a Unique Bond. Our Nation is fortunate to have the extraordinary power of the Navy on, above, and under the sea, coupled uniquely with the unmatched flexibility and ferocity of the U.S. Marine Corps. The character, structure, and tone of a force that define the Navy find like traits in the Marine Corps. It is not coincidental. It is a common bond forged by projecting power from the sea, a bond as old as the Nation and one to be nurtured as long as the Nation endures.

The U.S. Air Force, by General John P. Jumper, USAF (Ret.), Chief of Staff, 2001–2005

During the evening of March 27, 1999, while I was serving as Commander of United States Air Forces in Europe, the North Atlantic Treaty Organization (NATO) was fully engaged in Operation *Allied Force*, the alliance effort to expel Slobodan Milosevic's military forces from Kosovo. The evening did not start well. At approximately 2030 hours an F-117 stealth fighter was tracked and shot down by a Serbian SA-3 surface-to-air missile. In the real-time world of integrated command and control, our entire chain of command knew instantly as the siren sound of the ejection seat survival beacon echoed over the universal distress frequency, requiring no further explanation as to what had happened. In my command center, the hot lines erupted with demands to know how this could happen to a supposedly invisible stealth aircraft, and anxious discussions about the political consequences of an American pilot falling into Serbian hands. Air operations were assumed to be free of risk, somehow out of reach of the enemy. That was the low point. What unfolded throughout the rest of the evening was predictable, and confirmed my pride in serving our Nation as one of thc 300,000 men and women—active duty, Air National Guard, and Air Force Reserve—who call ourselves Airmen.

On that night, Airmen from around the globe would contribute their remarkable skills as the little-known and less-understood world of combat search and rescue (CSAR) came to life, awakened by the dreaded wail of a distress beacon. Within minutes, the Rescue Coordination and Air Operations Centers in Italy began to assemble the collection of capabilities required to bring this downed Airman home. Intelligence, surveillance, and electronic warfare platforms—airborne, manned and unmanned; satellites with very special capabilities for providing precise location and identification; special operations forces from the Army, Navy, and Air Force; dozens of air-refueling aircraft; airborne early warning; a global network of intelligence analysis and command and control—all fell in on this single problem. On CSAR alert that night, from a base in Italy, was a flight of four A-10 Warthogs led by a young captain. It could have been any of several flight-lead qualified captains in that squadron. It really didn't matter. But this captain was a graduate from the Air Force Weapons School. I did not know him at the time, but I was very familiar with the rigorous training he had endured to earn the Weapons School patch. This night he was the CSAR commander.

From our command center we listened as he led his flight toward enemy territory while organizing his globally sourced team: first, get a precise fix on the location; then, put the strike assets on the tanker to be ready when needed; move the tankers and surveillance assets in as close as possible to the surface-to-air missile rings; get the special operators airborne with the helicopter refueling assets; set up the combat air patrols and airborne early warning radars to deal with enemy fighters; position the ground-mapping radars to detect movement; activate communication networks to deal with enemy jamming; and complete the myriad tasks necessary to set the line and call the play.

All did not go well. There was initial confusion about the location of the downed Airman. One hour passed, then two, while the A-10s probed to test enemy reaction and reconcile inaccurate location data. Everyone's patience was tested, except for our captain who had organized his team to methodically work the problem. As patience thinned my hot lines grew hotter with suggestions that I step in personally to oversee the operation, that something different had to happen. My simple reply: "There is a captain commanding this rescue force, and no one on the planet is better trained or better positioned to complete this mission. We will let him do his job."

Then, we had radio contact with the downed Airman and an accurate location from our space forces. As the CSAR helicopters infiltrated with their special operations teams, we watched this armada, simple green blips on a screen moving toward the pickup point. Serbian ground forces were also moving closer to the pickup area. As the helicopters drew near, so did the rest of the dozens of aircraft stacked in orbits, ever closer to the scene, pressing inside of surface-to-air missile rings, whatever it took to get our guy. Ground fire and communications jamming were intense in the pickup area, but there was no enemy action to engage the many aircraft poised to engage when needed—no missiles firing, no fighters scrambling. The A-10s maneuvered to decoy the Serbian ground forces and then, with the downed pilot doing all the right things to guide the helicopters to his location, our special operations team rescued the Airman in sight of closing Serb ground forces. All safely returned to base.

I reflect back on this operation with the greatest pride. Within an hour, all of the technical skills, dedication, and energy of hundreds of Airmen were repurposed to create a team with singular, intense focus. The best training in the world taught them how to plan and execute this most difficult mission without notice or rehearsal. All grasped the urgency of the task and shared the determination to succeed. It didn't matter if you were a fighter pilot, a member of a tanker, space, or surveillance crew, or part of a special operations team. From the commanders in the operations centers to the hundreds of Airmen who maintain the platforms, load the weapons, create and maintain the networks, fly the satellites, control the airspace, work the complex mobility and logistics supporting many bases, each role was demonstrated and validated on this night. No one Airman could have been successful without the success of all other Airmen. And our captain? He was introduced by the President of the United States at the next State of the Union address and went on to become a general officer in the United States Air Force.

As Airmen we master the technology to control the speed and time compression of the vertical dimension in air and space. We train to rally our forces rapidly and globally. We make our Nation's joint forces better and coalition operations possible. We are mindful of our oath to defend the greatest Nation on earth and dedicate our service to its citizens.

The U.S. Coast Guard, by Admiral James M. Loy, USCG (Ret.), Commandant, 1998–2002

Close your eyes and place a finger anywhere on a map of the United States and you're likely to pinpoint a place where the United States Coast Guard is providing some sort of valuable service to the Nation and its citizens. Wherever you find cargo ships, sailboats, motorboats, bridges, ports of entry, or tankers full of crude oil, the Coast Guard is there. Wherever there is a problem in navigable U.S. waters with drugs, illegal migrants, customs issues, or smuggling, the Coast Guard is there. Whenever and wherever there is a need for homeland security or operations in defense of the United States, the Coast Guard is on hand performing its duty. The Coast Guard often operates without fanfare, routinely in support of other government agencies or military Services, frequently initiating or leading the action. And of course, all Americans know that when they are in trouble at sea, the Coast Guard will answer their distress call, promptly and efficiently.

The service's greatest strength is its multimission character wherein its agility and adaptability have become legendary. As one member of Congress once told me at a hearing, "If the mission is really hard and really wet, we're going to give it to the Coast Guard!"

So the inevitable question, of course, is: "How does the Coast Guard do so much so efficiently, with so few people, and so little money?"

First, it's the people. There are no spectators in the Coast Guard. Everybody performs several jobs, and the people match the multimission nature of the organization. The Coast Guard is filled with inspired and dedicated people of character and humility who do great things every day.

Second, the Coast Guard lives and breathes leadership. It pervades every aspect of an organization in which every person is a leader. One extraordinary example of that reality is the Coast Guard performance during Hurricane Katrina in New Orleans in 2005. When it appeared every other organization was failing in its responsibilities, the Coast Guard was doing almost 5 years' worth of Search and Rescue cases in 10 days. Those cases were not being done by admirals and captains; they were being done by young lieutenants with their hands on the controls of a helicopter, and by young petty officers and more junior enlisted personnel in flat-bottom boats who simply rose to the occasion and got done what needed to be done. So why was the performance standard for the Coast Guard so high? Why were the Coast Guard operations so ably led? Where does such quality come from?

The roots of the Coast Guard go back to the birth of the United States of America. It was a service organization imbued with proper leadership thinking and behavior by the Nation's founders. When our government was formed, President George Washington recognized that the next challenge was to build a new country. Of particular concern to the new President was the establishment of economic stability in the wake of the $70 million debt accumulated during the war. To take on this monumental task, President Washington chose Alexander Hamilton, his friend and former aide-de-camp who had fought alongside him from Trenton to Yorktown. Hamilton's new economic plan concentrated on shipping, which was then, and remains today, the world's primary mode of commerce and trade. He proposed customs duties and tariffs on imported goods, graduated tax rates on revenue, and various other shipping duties. He also advocated Federal responsibility for ensuring safety at sea for ships, their crews, and their cargoes under the American flag. To make such initiatives work, Hamilton believed that "a few armed vessels, judiciously stationed at the entrances of our ports might at a small expense, be made useful sentinels of the laws." Accordingly, he proposed the formation of a seagoing military force that would enforce customs and navigation laws, cruise the coasts, hail in-bound ships, make inspections, and certify manifests.

The first U.S. Congress formally approved Hamilton's proposal. The Act of August 4, 1790, provided for the establishment and support of 10 cutters, along with the creation of a professional corps of 40 commissioned officers to man the new service. The new cutter personnel were given the rank and standing of military officers because, as Hamilton said, "it would attach them to their duty by a nicer sense of honor." Hamilton penned the following letter to the new officers:

> *Always keep in mind that your countrymen are freemen, and as such, are impatient of everything that bears the least mark of a domineering spirit. . . . Refrain from haughtiness, rudeness, or insult. . . . Endeavor to overcome difficulties by a cool and*

temperate perseverance in your duty—by skill and moderation, rather than by vehemence or violence.

Hamilton further stated that his words of advice had been "selected with careful attention to character."[3]

That leadership standard has endured for more than two and one-quarter centuries, uncorrupted by the "business management" thinking of the industrial age, often tested by war, and recently in our time, tempered by terrorism in the homeland.

During my time as the Chief, Office of Personnel and Training at Coast Guard headquarters, I was tasked by the Commandant, Admiral Bill Kime, to produce a set of core values for the service. I assembled a task force of Coast Guard people from all aspects of the service—young and old, civilian and military, officers and enlisted, reserves and Coast Guard auxiliarists. We worked the project for several months, because I sensed the outcome had the potential to resonate far and wide. We grappled with many lists and checked out the core values of many organizations and military Services around the world. We conducted sessions at Coast Guard commands around the service. When I took the final product to Admiral Kime for his consideration and decision, he asked me how I would know if we got it right. Back we went to the mess decks and engine rooms of ships, to the hangar decks of the air stations, and to the lifeboat stations where Coast Guard people worked. We knew we had it right when our proposed set of core values resonated so well with real Coast Guard people. One of the most gratifying remarks I recall was from an E6 petty officer with 18 years in the Coast Guard. He said "Admiral, this set of core values could have been issued by Hamilton himself!"

So what are these core values? They are Honor, Respect, and Devotion to Duty—words that capture the culture of the Coast Guard as an organization and the mandates for every member.

Honor means high ethical conduct, moral behavior, honesty, integrity, trust, and doing what's right, not what's easy. It means honoring the traditions and the principles that make the Coast Guard and the United States what they are today.

Respect is one of the most important attributes of leadership. From respect spring caring, compassion, understanding, and effective communication. It is essentially the embodiment of the Golden Rule: Treat others as you would have them treat you. But, in the Coast Guard, it is a hard-and-fast rule. Everyone must treat others fairly and with civility, consideration, and dignity.

Devotion to Duty is the basic acceptance of responsibility, accountability, and commitment to doing your job. It's about taking pride in what you do. It's about a higher calling. "In our organization," says one admiral, "we exist to serve our country and its citizens. We serve with pride. We are devoted to preserving life at war and at home." Devotion to Duty tends to create an organization of doers. Coast Guard officers are committed to showing up for work on time and staying as long as is necessary to get the job done. They are always at their stations, always alert, and always ready to serve.

What does it mean to be a Coast Guard officer? It means that you work very hard to be ready to perform your duty every day. It means that the charge of Alexander Hamilton to the original 40 Revenue Marine officers is alive and well today. It means today's Coast Guard officer can not only recite the Coast Guard's Core Values, but can give you a full explanation as to what each one means. Coastguardsmen can do that because they live them every day.

Notes

[1] Letter dated February 12, 1918, from Brigadier General Charles A. Doyen, Headquarters, Fourth Brigade, Marine Corps, American Expeditionary Forces, to Brigadier General Charles Henry Lauchheimer, Headquarters Marine Corps. Letter provided by Dr. James Ginter, Archivist, U.S. Marine Corps History Office, Quantico, VA.

[2] Editorial, "Hobson's Choice," *Wall Street Journal*, May 14, 1952.

[3] Alexander Hamilton's Letter of Instructions to the Commanding Officers of the Revenue Cutters, Treasury Department, June 4, 1791, available at <www.uscg.mil/history/faqs/hamiltonletter.pdf>.

CHAPTER TEN

The Armed Forces Officer

"The choice of a line of work," states Professor William Lee Miller, "can be one of the foremost 'moral' choices one makes." It is, Miller continues, "a choice about what it is worthwhile to spend one's life doing."[1] The decision to undertake a military career of whatever duration, to accept an officer's commission, and to take the officer's oath is particularly weighty. It requires no less than commitment of one's life to the service of others. In exchange, such service carries with it the benefits and burdens of life as a public official in the world's most successful democracy and membership in an ancient and honorable calling—the profession of arms. Speaking of his own commission, George Washington wrote to a British opponent:

> *I cannot conceive of any more honorable* [source of officer's rank], *than that which flows from the uncorrupted Choice of a brave and free People—The purest Source & original Fountain of all Power.*[2]

As an American Armed Forces officer, one accepts responsibility both for faithful execution of the office, to include a life of continuous study and application, and for the maintenance of an exemplary personal life. This responsibility is owed to the Nation, fellow Armed Forces officers, all those who wear and have worn the Nation's uniform in any grade or capacity, as well as those who will come hereafter. The responsibility implies a dual obligation—to protect the Constitution and to pass on to others unsullied the honor of being an Armed Forces officer.

George Marshall was right: There is a common ground, ethically and morally binding all American military officers, of whatever service, to their particular branch and their fellow Armed Forces officers. This common ground originates with the common constitutional oath and commission. Indeed, it is the basis of the true professional *jointness* of the commissioned leaders of all the Armed Forces. Logically, it would be as true to say that all officers are commissioned into the Armed Forces of the United States, with service in a particular department, as it is to continue to follow the traditional form of commissioning them into the separate departments and binding them by a common oath and commission. In that sense, all officers are joint officers who happen to be on the rolls of their particular service. It is the common moral obligation that unites the separate service cultures into one fabric—*E pluribus unum*.

An officer of the Armed Forces of the United States must be a warrior, a leader of character, an unwavering defender of the Constitution, a servant of the Nation, and an exemplar and champion of its ideals.

Fighting, and leading those who do, is the unique role of Armed Forces officers. It is the warrior spirit that sustains men and women in times of danger, hardship, and discouragement, and that gives leaders the confidence and purpose to rally troops for one more effort when their will seems to be waning. According to Field Marshal Sir William Slim:

> *When you're in command and things have gone wrong there always comes a pause when your men stop and—they look at you. They don't say anything—they just look at you. It's rather an awful moment for the commander because then he knows that their courage is ebbing, their will is fading, and he's got to pull up out of himself the courage and the will power that will stiffen them again and make them go on.*[3]

Slim was reflecting on his role as an Army commander during the march out of Burma in War II, but the phenomenon applies even more surely to the platoon commander, division chief, or flight leader in the midst of battle. The warrior ethos is George Washington, who almost single-handedly sustained the Revolution by maintaining the will of the Continental army through his indomitable example in leading the attacks at Trenton and Princeton in the depths of the winter of 1776. It is Ulysses Grant at Fort Donelson, his line broken and troops driven back, riding to the front and telling his soldiers, "Fill your cartridge-boxes quick, and get into the line; the enemy is trying to escape, and he must not be permitted to do so."[4] It is Captain Guy V. Henry, lying wounded at the battle of the Rosebud during the Great Sioux War of 1976, telling a friend, "It is nothing. For this are we soldiers."[5] It is Admiral Chester Nimitz, ordering Admiral Raymond Spruance to be governed by the principle of calculated risk before the Battle of Midway, then sending him into battle against a superior Japanese fleet.[6] It is the indomitable spirit of Admiral James Stockdale, continuing to resist the Nation's enemies in spite of injury, captivity, and torture. Warriors will always have The Code of Conduct as their guide and standard: "*I am an American, fighting in the forces which guard my country and our way of life. I am prepared to give my life in their defense.*"

American warriors, of course, are not simply expected to win. They are expected to win constrained by the values cherished by the American people. The application of national values has changed over time, depending, among other things, on the nature of the war and the value of its objective to the American people. At a minimum, the American Armed Forces are expected to fight according to the principles of "Just War" enshrined in international conventions to which the Nation is a party. Violation of these rules, however inconvenient or dangerous those rules might be to one's self or one's unit, is contrary to U.S. law and indicative of a failure of professional discipline as well as of professional morality. This expectation of honorable arms is increasingly important as the actions of even the most junior troops become immediately visible to the world in an era of instantaneous communications. When the Armed Forces are functioning properly, everyone can expect that such violations will be prosecuted energetically.

Officers are expected to be leaders of character in peace as well as in war. Officers are creatures of the law, acting under authority of the President as constitutional Commander in Chief, according to the laws and regulations laid down by Congress. Because they are public figures entrusted with the means of war and authority over the lives of fellow citizens, officers' conduct must conform *at all times* to the highest standards of respect, honor, duty, service, integrity, excellence, courage, commitment, and loyalty. To do less undermines the credit of one's service, as well as the professional standing of the corps of American Armed Forces officers as public trustees of the Nation's welfare and security.

The Armed Forces officer, as a leader of character, is responsible not just for his or her own actions, but for protecting subordinates from the dehumanization that naturally follows descent into the maelstrom of war. The officer must stand above the chaos and travail and guard his or her people's humanity when it is most sorely tried. To do that, an officer must be very secure in the values the Nation and its armed services stand for and revere, in accordance with the special trust and confidence the President and the Nation have reposed in every officer's patriotism, valor, fidelity, and abilities.

The core of the Armed Forces officer's oath is to support and defend the Constitution, while bearing to it true faith and allegiance. Support and defense of the Constitution require, first of all, personal subordination to the civil officials established by the Constitution and the Congress to hold ultimate command on behalf of the American people. By their oaths, Armed Forces officers are co-opted for the duration of their commission to support and execute, even at the risk of their lives, the legal decisions of their civilian leaders, even when they believe they are ill-founded or ill-advised. When General Matthew Ridgway became Army Chief of Staff, he listed three primary responsibilities of the military professional:

> *First, to give his honest, fearless, objective, professional military opinion of what he needs to do the job the nation gives him. Second, if what he is given is less than the minimum he regards as essential, to give his superiors an honest, fearless, objective opinion of the consequences of these shortages as he sees them from the military viewpoint. Third and finally, he has the duty, whatever be the final decision, to do the utmost with whatever he is furnished.*[7]

Service to the Nation implies sustained preparation to deliver reliable and effective service on the day of battle. Armed Forces officers must continuously assess their technical skills, and those of their subordinates, and upgrade both by training, study, and practice. Officers must be imaginative, adaptive, and able to respond quickly to new circumstances and threats. They must be self-confident enough in their own skills and abilities to assume responsibility for taking action, even when out of sight and the immediate control of superiors. They must be self-aware, self-reflective, and self-critical. The American people entrust their sons and daughters to officers' care. For all of these reasons, competence in every aspect of the profession of arms is a moral obligation.

Finally, Armed Forces officers are expected to reflect the Nation's ideals in all that they do. Sadly, the conduct of military professionals will not always be up to standard. Every member of the profession of arms has an obligation to do something to address perceived failures, by questioning, by encouraging, and in egregious cases, by being willing to act. "If you see something, say something," or even better, *do something.* Putting on blinders to the misconduct of others, or being passive in the face of violations by others, is a failure to fulfill a solemn obligation to the institution, to the profession, to this ancient and honorable calling. Every officer is responsible for his or her own conduct. Beyond that, every officer is responsible for ensuring that the standards of the profession are upheld, practiced, and enforced by all its members, whether junior, peer, or senior. The higher the rank, the greater is that obligation.

Every officer must have a moral compass, and periodically recalibrate it to ensure that it is still pointing to true ethical north. The standard is always what is good for the Nation, not what is good in the short term for the profession or the particular armed Service. Narrow loyalty to the latter can lead to individual and collective deceptions that, in the end, are corrosive of the honor of the profession and all its members. What is good for America is always good for the Armed Forces.

Armed Forces officers carry on an enduring tradition of citizen service to the Nation. Their conduct must honor the ideals and principles enshrined in the Declaration of Independence: *that all men are created equal, that they are endowed by their Creator with certain unalienable Rights, that among these are Life, Liberty, and the pursuit of Happiness.* The officer's demonstrated character, marked by integrity, courage, capability, and commitment, must be such that he or she is worthy of following into harm's way. The officer as a public figure must model values of a higher standard than those often celebrated in the popular culture, and they must do so without succumbing to the conceit of believing they are better than their masters, the American people:

> *Only when the military articulates and lives up to its highest values can it retain the nobility of the profession of arms. Only when it retains a proper sense of its role in American democratic life does it retain the trust and respect [George C.] Marshall spoke of. Only a military that daily lives out its values and feels its connection to the citizens is a military that engenders the respect and loyalty of the nation and keeps it from being feared.*[8]

Every American Armed Forces officer has entered an ancient and honorable calling, a life of discipline, hardship, and danger. It is, therefore, a heroic life.[9] At the end of an officer's service, no matter how short or long, the reward will be the satisfaction of knowing that character, competence, and leadership made a difference in his or her own life, the lives of troops led, and the lives of fellow citizens.

Notes

[1] William Lee Miller, *Lincoln's Virtues* (New York: Vintage Books, 2003), 92.

[2] George Washington to Thomas Gage, August 19, 1775, in George Washington, *George Washington: Writings*, ed. John H. Rhodehamel (New York: Library of America, 1997), 182.

[3] Field Marshal Sir William Slim, "Higher Command in War," 1952 Kermit Roosevelt Lecture, audio tape, Combined Arms Research Library, U.S. Army Command and General Staff College, Fort Leavenworth, KS.

[4] Quoted in J.F.C. Fuller, *The Generalship of Ulysses S. Grant* (Bloomington: Indiana University Press, 1977), 88.

[5] John F. Finerty, *Warpath and Bivouac* (Chicago: M.A. Donohue & Co., 1890), 130.

[6] Thomas B. Buell, *The Quiet Warrior: A Biography of Admiral Raymond A. Spruance* (Boston: Little Brown, 1974), 123–124.

[7] Matthew Ridgway, *Soldier: The Memoirs of Matthew B. Ridgway as Told to Harold H. Martin* (New York: Harper, 1956), 346.

[8] Martin L. Cook, "Moral Foundations of Military Service," *Parameters* 30, no. 1 (Spring 2000), 117–129, available at <http://carlisle-www.army.mil/usawc/Parameters/00spring/cook.htm>.

[9] Jules J. Toner, S.J., quoted in Chris Lowney, *Heroic Leadership: Best Practices From a 450-year-old Company that Changed the World* (Chicago: Loyola Press, 2003), 49. What was written about a religious order applies equally well to the profession of arms: "one who truly lives under obedience is fully disposed to execute instantly and unhesitatingly whatever is enjoined him [or her], no matter whether it be very hard to do."

Appendix A: Founding Documents

The Declaration of Independence

IN CONGRESS, July 4, 1776.

The unanimous Declaration of the thirteen united States of America,

When in the Course of human events, it becomes necessary for one people to dissolve the political bands which have connected them with another, and to assume among the powers of the earth, the separate and equal station to which the Laws of Nature and of Nature's God entitle them, a decent respect to the opinions of mankind requires that they should declare the causes which impel them to the separation.

We hold these truths to be self-evident, that all men are created equal, that they are endowed by their Creator with certain unalienable Rights, that among these are Life, Liberty and the pursuit of Happiness.—That to secure these rights, Governments are instituted among Men, deriving their just powers from the consent of the governed, —That whenever any Form of Government becomes destructive of these ends, it is the Right of the People to alter or to abolish it, and to institute new Government, laying its foundation on such principles and organizing its powers in such form, as to them shall seem most likely to effect their Safety and Happiness. Prudence, indeed, will dictate that Governments long established should not be changed for light and transient causes; and accordingly all experience hath shewn, that mankind are more disposed to suffer, while evils are sufferable, than to right themselves by abolishing the forms to which they are accustomed. But when a long train of abuses and usurpations, pursuing invariably the same Object evinces a design to reduce them under absolute Despotism, it is their right, it is their duty, to throw off such Government, and to provide new Guards for their future security.—Such has been the patient sufferance of these Colonies; and such is now the necessity which constrains them to alter their former Systems of Government. The history of the present King of Great Britain is a history of repeated injuries and usurpations, all having in direct object the establishment of an absolute Tyranny over these States. To prove this, let Facts be submitted to a candid world.

He has refused his Assent to Laws, the most wholesome and necessary for the public good.

He has forbidden his Governors to pass Laws of immediate and pressing importance, unless suspended in their operation till his Assent should be obtained; and when so suspended, he has utterly neglected to attend to them.

He has refused to pass other Laws for the accommodation of large districts of people, unless those people would relinquish the right of Representation in the Legislature, a right inestimable to them and formidable to tyrants only.

He has called together legislative bodies at places unusual, uncomfortable, and distant from the depository of their public Records, for the sole purpose of fatiguing them into compliance with his measures.

He has dissolved Representative Houses repeatedly, for opposing with manly firmness his invasions on the rights of the people.

He has refused for a long time, after such dissolutions, to cause others to be elected; whereby the Legislative powers, incapable of Annihilation, have returned to the People at large for their exercise; the State remaining in the mean time exposed to all the dangers of invasion from without, and convulsions within.

He has endeavoured to prevent the population of these States; for that purpose obstructing the Laws for Naturalization of Foreigners; refusing to pass others to encourage their migrations hither, and raising the conditions of new Appropriations of Lands.

He has obstructed the Administration of Justice, by refusing his Assent to Laws for establishing Judiciary powers.

He has made Judges dependent on his Will alone, for the tenure of their offices, and the amount and payment of their salaries.

He has erected a multitude of New Offices, and sent hither swarms of Officers to harrass our people, and eat out their substance.

He has kept among us, in times of peace, Standing Armies without the Consent of our legislatures.

He has affected to render the Military independent of and superior to the Civil power.

He has combined with others to subject us to a jurisdiction foreign to our constitution, and unacknowledged by our laws; giving his Assent to their Acts of pretended Legislation:

For Quartering large bodies of armed troops among us:

For protecting them, by a mock Trial, from punishment for any Murders which they should commit on the Inhabitants of these States:

For cutting off our Trade with all parts of the world:

For imposing Taxes on us without our Consent:

For depriving us in many cases, of the benefits of Trial by Jury:

For transporting us beyond Seas to be tried for pretended offences

For abolishing the free System of English Laws in a neighbouring Province, establishing therein an Arbitrary government, and enlarging its Boundaries so as to render it at once an example and fit instrument for introducing the same absolute rule into these Colonies:

For taking away our Charters, abolishing our most valuable Laws, and altering fundamentally the Forms of our Governments:

For suspending our own Legislatures, and declaring themselves invested with power to legislate for us in all cases whatsoever.

He has abdicated Government here, by declaring us out of his Protection and waging War against us.

He has plundered our seas, ravaged our Coasts, burnt our towns, and destroyed the lives of our people.

He is at this time transporting large Armies of foreign Mercenaries to compleat the works of death, desolation and tyranny, already begun with circumstances of Cruelty & perfidy scarcely paralleled in the most barbarous ages, and totally unworthy the Head of a civilized nation.

He has constrained our fellow Citizens taken Captive on the high Seas to bear Arms against their Country, to become the executioners of their friends and Brethren, or to fall themselves by their Hands.

He has excited domestic insurrections amongst us, and has endeavoured to bring on the inhabitants of our frontiers, the merciless Indian Savages, whose known rule of warfare, is an undistinguished destruction of all ages, sexes and conditions.

In every stage of these Oppressions We have Petitioned for Redress in the most humble terms: Our repeated Petitions have been answered only by repeated injury. A Prince whose character is thus marked by every act which may define a Tyrant, is unfit to be the ruler of a free people.

Nor have We been wanting in attentions to our Brittish brethren. We have warned them from time to time of attempts by their legislature to extend an unwarrantable jurisdiction over us. We have reminded them of the circumstances of our emigration and settlement here. We have appealed to their native justice and magnanimity, and we have conjured them by the ties of our common kindred to disavow these usurpations, which, would inevitably interrupt our connections and correspondence. They too have been deaf to the voice of justice and of consanguinity. We must, therefore, acquiesce in the necessity, which denounces our Separation, and hold them, as we hold the rest of mankind, Enemies in War, in Peace Friends.

We, therefore, the Representatives of the united States of America, in General Congress, Assembled, appealing to the Supreme Judge of the world for the rectitude of our intentions, do, in the Name, and by Authority of the good People of these Colonies, solemnly publish and declare, That these United Colonies are, and of Right ought to be Free and Independent States; that they are Absolved from all Allegiance to the British Crown, and that all political connection between them and the State of Great Britain, is and ought to be totally dissolved; and that as Free and Independent States, they have full Power to levy War, conclude Peace, contract Alliances, establish Commerce, and to do all other Acts and Things which Independent States may of right do. And for the support of this Declaration, with a firm reliance on the protection of divine Providence, we mutually pledge to each other our Lives, our Fortunes and our sacred Honor.

The 56 signatures on the Declaration appear in the positions indicated:

Column 1

Georgia:

Button Gwinnett
Lyman Hall
George Walton

Column 2

North Carolina:

William Hooper
Joseph Hewes
John Penn

South Carolina:

Edward Rutledge
Thomas Heyward, Jr.
Thomas Lynch, Jr.
Arthur Middleton

Column 3

Massachusetts:

John Hancock

Maryland:

Samuel Chase
William Paca
Thomas Stone
Charles Carroll of Carrollton

Virginia:

George Wythe
Richard Henry Lee
Thomas Jefferson
Benjamin Harrison
Thomas Nelson, Jr.
Francis Lightfoot Lee
Carter Braxton

Column 4

Pennsylvania:

Robert Morris
Benjamin Rush
Benjamin Franklin
John Morton
George Clymer
James Smith
George Taylor
James Wilson
George Ross

Delaware:

Caesar Rodney
George Read
Thomas McKean

Column 5

New York:

William Floyd
Philip Livingston
Francis Lewis
Lewis Morris

New Jersey:

Richard Stockton
John Witherspoon
Francis Hopkinson
John Hart
Abraham Clark

Column 6

New Hampshire:

Josiah Bartlett
William Whipple

Massachusetts:
Samuel Adams
John Adams
Robert Treat Paine
Elbridge Gerry
Rhode Island:
Stephen Hopkins
William Ellery
Connecticut:
Roger Sherman
Samuel Huntington
William Williams
Oliver Wolcott
New Hampshire:
Matthew Thornton

The Constitution of the United States

Note: The following text is a transcription of the Constitution in its original form.

Items that are underlined have since been amended or superseded.

We the People of the United States, in Order to form a more perfect Union, establish Justice, insure domestic Tranquility, provide for the common defense, promote the general Welfare, and secure the Blessings of Liberty to ourselves and our Posterity, do ordain and establish this Constitution for the United States of America.

Article. I.

Section. 1.

All legislative Powers herein granted shall be vested in a Congress of the United States, which shall consist of a Senate and House of Representatives.

Section. 2.

The House of Representatives shall be composed of Members chosen every second Year by the People of the several States, and the Electors in each State shall have the Qualifications requisite for Electors of the most numerous Branch of the State Legislature.

No Person shall be a Representative who shall not have attained to the Age of twenty five Years, and been seven Years a Citizen of the United States, and who shall not, when elected, be an Inhabitant of that State in which he shall be chosen.

Representatives and direct Taxes shall be apportioned among the several States which may be included within this Union, according to their respective Numbers, which shall be determined by adding to the whole Number of free Persons, including those bound to Service for a Term of Years, and excluding Indians not taxed, three fifths of all other Persons. The actual Enumeration shall be made within three Years after the first Meeting of the Congress of the United States, and within every subsequent Term of ten Years, in such Manner as they shall by Law direct. The Number of Representatives shall not exceed one for every thirty Thousand, but each State shall have at Least one Representative; and until such enumeration shall be made, the State of New Hampshire shall be entitled to chuse three, Massachusetts eight, Rhode-Island and Providence Plantations one, Connecticut five, New-York six, New Jersey four, Pennsylvania eight, Delaware one, Maryland six, Virginia ten, North Carolina five, South Carolina five, and Georgia three.

When vacancies happen in the Representation from any State, the Executive Authority thereof shall issue Writs of Election to fill such Vacancies.

The House of Representatives shall chuse their Speaker and other Officers; and shall have the sole Power of Impeachment.

Section. 3.

The Senate of the United States shall be composed of two Senators from each State, chosen by the Legislature thereof for six Years; and each Senator shall have one Vote.

Immediately after they shall be assembled in Consequence of the first Election, they shall be divided as equally as may be into three Classes. The Seats of the Senators of the first Class shall be vacated at the Expiration of the second Year, of the second Class at the Expiration of the fourth Year, and of the third Class at the Expiration of the sixth Year, so that one third may be chosen every second Year; and if Vacancies happen by Resignation, or otherwise, during the Recess of the Legislature of any State, the Executive thereof may make temporary Appointments until the next Meeting of the Legislature, which shall then fill such Vacancies.

No Person shall be a Senator who shall not have attained to the Age of thirty Years, and been nine Years a Citizen of the United States, and who shall not, when elected, be an Inhabitant of that State for which he shall be chosen.

The Vice President of the United States shall be President of the Senate, but shall have no Vote, unless they be equally divided.

The Senate shall chuse their other Officers, and also a President pro tempore, in the Absence of the Vice President, or when he shall exercise the Office of President of the United States.

The Senate shall have the sole Power to try all Impeachments. When sitting for that Purpose, they shall be on Oath or Affirmation. When the

President of the United States is tried, the Chief Justice shall preside: And no Person shall be convicted without the Concurrence of two thirds of the Members present.

Judgment in Cases of Impeachment shall not extend further than to removal from Office, and disqualification to hold and enjoy any Office of honor, Trust or Profit under the United States: but the Party convicted shall nevertheless be liable and subject to Indictment, Trial, Judgment and Punishment, according to Law.

Section. 4.

The Times, Places and Manner of holding Elections for Senators and Representatives, shall be prescribed in each State by the Legislature thereof; but the Congress may at any time by Law make or alter such Regulations, except as to the Places of chusing Senators.

The Congress shall assemble at least once in every Year, and such Meeting shall be on the first Monday in December, unless they shall by Law appoint a different Day.

Section. 5.

Each House shall be the Judge of the Elections, Returns and Qualifications of its own Members, and a Majority of each shall constitute a Quorum to do Business; but a smaller Number may adjourn from day to day, and may be authorized to compel the Attendance of absent Members, in such Manner, and under such Penalties as each House may provide.

Each House may determine the Rules of its Proceedings, punish its Members for disorderly Behaviour, and, with the Concurrence of two thirds, expel a Member.

Each House shall keep a Journal of its Proceedings, and from time to time publish the same, excepting such Parts as may in their Judgment require Secrecy; and the Yeas and Nays of the Members of either House on any question shall, at the Desire of one fifth of those Present, be entered on the Journal.

Neither House, during the Session of Congress, shall, without the Consent of the other, adjourn for more than three days, nor to any other Place than that in which the two Houses shall be sitting.

Section. 6.

The Senators and Representatives shall receive a Compensation for their Services, to be ascertained by Law, and paid out of the Treasury of the United States. They shall in all Cases, except Treason, Felony and Breach of the Peace, be privileged from Arrest during their Attendance at the Session of their respective Houses, and in going to and returning from the same; and for any Speech or Debate in either House, they shall not be questioned in any other Place.

No Senator or Representative shall, during the Time for which he was elected, be appointed to any civil Office under the Authority of the United States, which shall have been created, or the Emoluments whereof shall have been encreased during such time; and no Person holding any Office under the United States, shall be a Member of either House during his Continuance in Office.

Section. 7.

All Bills for raising Revenue shall originate in the House of Representatives; but the Senate may propose or concur with Amendments as on other Bills.

Every Bill which shall have passed the House of Representatives and the Senate, shall, before it become a Law, be presented to the President of the United States: If he approve he shall sign it, but if not he shall return it, with his Objections to that House in which it shall have originated, who shall enter the Objections at large on their Journal, and proceed to reconsider it. If after such Reconsideration two thirds of that House shall agree to pass the Bill, it shall be sent, together with the Objections, to the other House, by which it shall likewise be reconsidered, and if approved by two thirds of that House, it shall become a Law. But in all such Cases the Votes of both Houses shall be determined by yeas and Nays, and the Names of the Persons voting for and against the Bill shall be entered on the Journal of each House respectively. If any Bill shall not be returned by the President within ten Days (Sundays excepted) after it shall have been presented to him, the Same shall be a Law, in like Manner as if he had signed it, unless the Congress by their Adjournment prevent its Return, in which Case it shall not be a Law.

Every Order, Resolution, or Vote to which the Concurrence of the Senate and House of Representatives may be necessary (except on a question of Adjournment) shall be presented to the President of the United States; and before the Same shall take Effect, shall be approved by him, or being disapproved by him, shall be repassed by two thirds of the Senate and House of Representatives, according to the Rules and Limitations prescribed in the Case of a Bill.

Section. 8.

The Congress shall have Power To lay and collect Taxes, Duties, Imposts and Excises, to pay the Debts and provide for the common Defence and general Welfare of the United States; but all Duties, Imposts and Excises shall be uniform throughout the United States;

To borrow Money on the credit of the United States;

To regulate Commerce with foreign Nations, and among the several States, and with the Indian Tribes;

To establish an uniform Rule of Naturalization, and uniform Laws on the subject of Bankruptcies throughout the United States;

To coin Money, regulate the Value thereof, and of foreign Coin, and fix the Standard of Weights and Measures;

To provide for the Punishment of counterfeiting the Securities and current Coin of the United States;

To establish Post Offices and post Roads;

To promote the Progress of Science and useful Arts, by securing for limited Times to Authors and Inventors the exclusive Right to their respective Writings and Discoveries;

To constitute Tribunals inferior to the supreme Court;

To define and punish Piracies and Felonies committed on the high Seas, and Offences against the Law of Nations;

To declare War, grant Letters of Marque and Reprisal, and make Rules concerning Captures on Land and Water;

To raise and support Armies, but no Appropriation of Money to that Use shall be for a longer Term than two Years;

To provide and maintain a Navy;

To make Rules for the Government and Regulation of the land and naval Forces;

To provide for calling forth the Militia to execute the Laws of the Union, suppress Insurrections and repel Invasions;

To provide for organizing, arming, and disciplining, the Militia, and for governing such Part of them as may be employed in the Service of the United States, reserving to the States respectively, the Appointment of the Officers, and the Authority of training the Militia according to the discipline prescribed by Congress;

To exercise exclusive Legislation in all Cases whatsoever, over such District (not exceeding ten Miles square) as may, by Cession of particular States, and the Acceptance of Congress, become the Seat of the Government of the United States, and to exercise like Authority over all Places purchased by the Consent of the Legislature of the State in which the Same shall be, for the Erection of Forts, Magazines, Arsenals, dock-Yards, and other needful Buildings;—And

To make all Laws which shall be necessary and proper for carrying into Execution the foregoing Powers, and all other Powers vested by this Constitution in the Government of the United States, or in any Department or Officer thereof.

Section. 9.

The Migration or Importation of such Persons as any of the States now existing shall think proper to admit, shall not be prohibited by the Congress prior to the Year one thousand eight hundred and eight, but a Tax or duty may be imposed on such Importation, not exceeding ten dollars for each Person.

The Privilege of the Writ of Habeas Corpus shall not be suspended, unless when in Cases of Rebellion or Invasion the public Safety may require it.

No Bill of Attainder or ex post facto Law shall be passed.

No Capitation, or other direct, Tax shall be laid, unless in Proportion to the Census or enumeration herein before directed to be taken.

No Tax or Duty shall be laid on Articles exported from any State.

No Preference shall be given by any Regulation of Commerce or Revenue to the Ports of one State over those of another; nor shall Vessels bound to, or from, one State, be obliged to enter, clear, or pay Duties in another.

No Money shall be drawn from the Treasury, but in Consequence of Appropriations made by Law; and a regular Statement and Account of the Receipts and Expenditures of all public Money shall be published from time to time.

No Title of Nobility shall be granted by the United States: And no Person holding any Office of Profit or Trust under them, shall, without the Consent of the Congress, accept of any present, Emolument, Office, or Title, of any kind whatever, from any King, Prince, or foreign State.

Section. 10.

No State shall enter into any Treaty, Alliance, or Confederation; grant Letters of Marque and Reprisal; coin Money; emit Bills of Credit; make any Thing but gold and silver Coin a Tender in Payment of Debts; pass any Bill of Attainder, ex post facto Law, or Law impairing the Obligation of Contracts, or grant any Title of Nobility.

No State shall, without the Consent of the Congress, lay any Imposts or Duties on Imports or Exports, except what may be absolutely necessary for executing it's inspection Laws: and the net Produce of all Duties and Imposts, laid by any State on Imports or Exports, shall be for the Use of the Treasury of the United States; and all such Laws shall be subject to the Revision and Controul of the Congress.

No State shall, without the Consent of Congress, lay any Duty of Tonnage, keep Troops, or Ships of War in time of Peace, enter into any Agreement or Compact with another State, or with a foreign Power, or engage in War, unless actually invaded, or in such imminent Danger as will not admit of delay.

Article. II.

Section. 1.

The executive Power shall be vested in a President of the United States of America. He shall hold his Office during the Term of four Years, and, together with the Vice President, chosen for the same Term, be elected, as follows:

Each State shall appoint, in such Manner as the Legislature thereof may direct, a Number of Electors, equal to the whole Number of Senators and Representatives to which the State may be entitled in the Congress: but no Senator or Representative, or Person holding an Office of Trust or Profit under the United States, shall be appointed an Elector.

The Electors shall meet in their respective States, and vote by Ballot for two Persons, of whom one at least shall not be an Inhabitant of the same State with themselves. And they shall make a List of all the Persons voted for, and of the Number of Votes for each; which List they shall sign and certify, and transmit sealed to the Seat of the Government of the United States, directed to the President of the Senate. The President of the Senate shall, in the Presence of the Senate and House of Representatives, open all the Certificates, and the Votes shall then be counted. The Person having the greatest Number of Votes shall be the President, if such Number be a Majority of the whole Number of Electors appointed; and if there be more than one who have such Majority, and have an equal Number of Votes, then the House of Representatives shall immediately chuse by Ballot one of them for President; and if no Person have a Majority, then from the five highest on the List the said House shall in like Manner chuse the President. But in chusing the President, the Votes shall be taken by States, the Representation from each State having one Vote; A quorum for this purpose shall consist of a Member or Members from two thirds of the States, and a Majority of all the States shall be necessary to a Choice. In every Case, after the Choice of the President, the Person having the greatest Number of Votes of the Electors shall be the Vice President. But if there should remain two or more who have equal Votes, the Senate shall chuse from them by Ballot the Vice President.

The Congress may determine the Time of chusing the Electors, and the Day on which they shall give their Votes; which Day shall be the same throughout the United States.

No Person except a natural born Citizen, or a Citizen of the United States, at the time of the Adoption of this Constitution, shall be eligible to the Office of President; neither shall any Person be eligible to that Office who shall not have attained to the Age of thirty five Years, and been fourteen Years a Resident within the United States.

In Case of the Removal of the President from Office, or of his Death, Resignation, or Inability to discharge the Powers and Duties of the said Office, the Same shall devolve on the Vice President, and the Congress may by Law provide for the Case of Removal, Death, Resignation or Inability, both of the President and Vice President, declaring what Officer shall then act as President, and such Officer shall act accordingly, until the Disability be removed, or a President shall be elected.

The President shall, at stated Times, receive for his Services, a Compensation, which shall neither be increased nor diminished during the Period for which he shall have been elected, and he shall not receive within that Period any other Emolument from the United States, or any of them.

Before he enter on the Execution of his Office, he shall take the following Oath or Affirmation:—"I do solemnly swear (or affirm) that I will faithfully execute the Office of President of the United States, and will to the best of my Ability, preserve, protect and defend the Constitution of the United States."

Section. 2.

The President shall be Commander in Chief of the Army and Navy of the United States, and of the Militia of the several States, when called into the actual Service of the United States; he may require the Opinion, in writing, of the principal Officer in each of the executive Departments, upon any Subject relating to the Duties of their respective Offices, and he shall have Power to grant Reprieves and Pardons for Offences against the United States, except in Cases of Impeachment.

He shall have Power, by and with the Advice and Consent of the Senate, to make Treaties, provided two thirds of the Senators present concur; and he shall nominate, and by and with the Advice and Consent of the Senate, shall appoint Ambassadors, other public Ministers and Consuls, Judges of the supreme Court, and all other Officers of the United States, whose Appointments are not herein otherwise provided for, and which shall be established by Law: but the Congress may by Law vest the Appointment of such inferior Officers, as they think proper, in the President alone, in the Courts of Law, or in the Heads of Departments.

The President shall have Power to fill up all Vacancies that may happen during the Recess of the Senate, by granting Commissions which shall expire at the End of their next Session.

Section. 3.

He shall from time to time give to the Congress Information of the State of the Union, and recommend to their Consideration such Measures as he shall judge necessary and expedient; he may, on extraordinary Occasions, convene both Houses, or either of them, and in Case of

Disagreement between them, with Respect to the Time of Adjournment, he may adjourn them to such Time as he shall think proper; he shall receive Ambassadors and other public Ministers; he shall take Care that the Laws be faithfully executed, and shall Commission all the Officers of the United States.

Section. 4.

The President, Vice President and all civil Officers of the United States, shall be removed from Office on Impeachment for, and Conviction of, Treason, Bribery, or other high Crimes and Misdemeanors.

ARTICLE III.

Section. 1.

The judicial Power of the United States shall be vested in one supreme Court, and in such inferior Courts as the Congress may from time to time ordain and establish. The Judges, both of the supreme and inferior Courts, shall hold their Offices during good Behaviour, and shall, at stated Times, receive for their Services a Compensation, which shall not be diminished during their Continuance in Office.

Section. 2.

The judicial Power shall extend to all Cases, in Law and Equity, arising under this Constitution, the Laws of the United States, and Treaties made, or which shall be made, under their Authority;—to all Cases affecting Ambassadors, other public Ministers and Consuls;—to all Cases of admiralty and maritime Jurisdiction;—to Controversies to which the United States shall be a Party;—to Controversies between two or more States;— between a State and Citizens of another State;—between Citizens of different States;—between Citizens of the same State claiming Lands under Grants of different States, and between a State, or the Citizens thereof, and foreign States, Citizens or Subjects.

In all Cases affecting Ambassadors, other public Ministers and Consuls, and those in which a State shall be Party, the supreme Court shall have original Jurisdiction. In all the other Cases before mentioned, the supreme Court shall have appellate Jurisdiction, both as to Law and Fact, with such Exceptions, and under such Regulations as the Congress shall make.

The Trial of all Crimes, except in Cases of Impeachment, shall be by Jury; and such Trial shall be held in the State where the said Crimes shall have been committed; but when not committed within any State, the Trial shall be at such Place or Places as the Congress may by Law have directed.

Section. 3.

Treason against the United States, shall consist only in levying War against them, or in adhering to their Enemies, giving them Aid and Comfort. No Person shall be convicted of Treason unless on the Testimony of two Witnesses to the same overt Act, or on Confession in open Court.

The Congress shall have Power to declare the Punishment of Treason, but no Attainder of Treason shall work Corruption of Blood, or Forfeiture except during the Life of the Person attainted.

ARTICLE. IV.

Section. 1.

Full Faith and Credit shall be given in each State to the public Acts, Records, and judicial Proceedings of every other State. And the Congress may by general Laws prescribe the Manner in which such Acts, Records and Proceedings shall be proved, and the Effect thereof.

Section. 2.

The Citizens of each State shall be entitled to all Privileges and Immunities of Citizens in the several States.

A Person charged in any State with Treason, Felony, or other Crime, who shall flee from Justice, and be found in another State, shall on Demand of the executive Authority of the State from which he fled, be delivered up, to be removed to the State having Jurisdiction of the Crime.

No Person held to Service or Labour in one State, under the Laws thereof, escaping into another, shall, in Consequence of any Law or Regulation therein, be discharged from such Service or Labour, but shall be delivered up on Claim of the Party to whom such Service or Labour may be due.

Section. 3.

New States may be admitted by the Congress into this Union; but no new State shall be formed or erected within the Jurisdiction of any other State; nor any State be formed by the Junction of two or more States, or Parts of States, without the Consent of the Legislatures of the States concerned as well as of the Congress.

The Congress shall have Power to dispose of and make all needful Rules and Regulations respecting the Territory or other Property belonging to the United States; and nothing in this Constitution shall be so construed as to Prejudice any Claims of the United States, or of any particular State.

Section. 4.

The United States shall guarantee to every State in this Union a Republican Form of Government, and shall protect each of them against Invasion; and on Application of the Legislature, or of the Executive (when the Legislature cannot be convened), against domestic Violence.

ARTICLE. V.

The Congress, whenever two thirds of both Houses shall deem it necessary, shall propose Amendments to this Constitution, or, on the Application of the Legislatures of two thirds of the several States, shall call a Convention for proposing Amendments, which, in either Case, shall be valid to all Intents and Purposes, as Part of this Constitution, when ratified by the Legislatures of three fourths of the several States, or by Conventions in three fourths thereof, as the one or the other Mode of Ratification may be proposed by the Congress; Provided that no Amendment which may be made prior to the Year One thousand eight hundred and eight shall in any Manner affect the first and fourth Clauses in the Ninth Section of the first Article; and that no State, without its Consent, shall be deprived of its equal Suffrage in the Senate.

ARTICLE. VI.

All Debts contracted and Engagements entered into, before the Adoption of this Constitution, shall be as valid against the United States under this Constitution, as under the Confederation.

This Constitution, and the Laws of the United States which shall be made in Pursuance thereof; and all Treaties made, or which shall be made, under the Authority of the United States, shall be the supreme Law of the Land; and the Judges in every State shall be bound thereby, any Thing in the Constitution or Laws of any State to the Contrary notwithstanding.

The Senators and Representatives before mentioned, and the Members of the several State Legislatures, and all executive and judicial Officers, both of the United States and of the several States, shall be bound by Oath or Affirmation, to support this Constitution; but no religious Test shall ever be required as a Qualification to any Office or public Trust under the United States.

Article. VII.

The Ratification of the Conventions of nine States, shall be sufficient for the Establishment of this Constitution between the States so ratifying the Same.

The Word, "the," being interlined between the seventh and eighth Lines of the first Page, the Word "Thirty" being partly written on an Erazure in the fifteenth Line of the first Page, The Words "is tried" being interlined between the thirty second and thirty third Lines of the first Page and the Word "the" being interlined between the forty third and forty fourth Lines of the second Page.

Attest William Jackson Secretary

Done in Convention by the Unanimous Consent of the States present the Seventeenth Day of September in the Year of our Lord one thousand seven hundred and Eighty seven and of the Independence of the United States of America the Twelfth In witness whereof We have hereunto subscribed our Names,

Go. Washington
Presidt and deputy from
Virginia

Delaware
Geo: Read
Gunning Bedford jun
John Dickinson
Richard Bassett
Jaco: Broom

Maryland
James McHenry
Dan of St Thos. Jenifer
Danl. Carroll

Virginia
John Blair
James Madison Jr.

North Carolina
Wm. Blount
Richd. Dobbs Spaight
Hu Williamson

South Carolina
J. Rutledge
Charles Cotesworth Pinckney
Charles Pinckney
Pierce Butler

Georgia
William Few
Abr Baldwin

New Hampshire
John Langdon
Nicholas Gilman

Massachusetts
Nathaniel Gorham
Rufus King

Connecticut
Wm. Saml. Johnson
Roger Sherman

New York
Alexander Hamilton

New Jersey
Wil: Livingston
David Brearley
Wm. Paterson
Jona: Dayton

Pennsylvania
B Franklin
Thomas Mifflin
Robt. Morris
Geo. Clymer
Thos. FitzSimons
Jared Ingersoll
James Wilson
Gouv Morris

The Bill of Rights

Note: The following text is a transcription of the first ten amendments to the Constitution in their original form. These amendments were ratified December 15, 1791, and form what is known as the "Bill of Rights."

Amendment I

Congress shall make no law respecting an establishment of religion, or prohibiting the free exercise thereof; or abridging the freedom of speech, or of the press; or the right of the people peaceably to assemble, and to petition the Government for a redress of grievances.

Amendment II

A well regulated Militia, being necessary to the security of a free State, the right of the people to keep and bear Arms, shall not be infringed.

Amendment III

No Soldier shall, in time of peace be quartered in any house, without the consent of the Owner, nor in time of war, but in a manner to be prescribed by law.

Amendment IV

The right of the people to be secure in their persons, houses, papers, and effects, against unreasonable searches and seizures, shall not be violated, and no Warrants shall issue, but upon probable cause, supported by Oath or affirmation, and particularly describing the place to be searched, and the persons or things to be seized.

Amendment V

No person shall be held to answer for a capital, or otherwise infamous crime, unless on a presentment or indictment of a Grand Jury, except in cases arising in the land or naval forces, or in the Militia, when in actual service in time of War or public danger; nor shall any person be subject for the same offence to be twice put in jeopardy of life or limb; nor shall be compelled in any criminal case to be a witness against himself, nor be deprived of life, liberty, or property, without due process of law; nor shall private property be taken for public use, without just compensation.

Amendment VI

In all criminal prosecutions, the accused shall enjoy the right to a speedy and public trial, by an impartial jury of the State and district wherein the crime shall have been committed, which district shall have been previously ascertained by law, and to be informed of the nature and cause of the accusation; to be confronted with the witnesses against him; to have compulsory process for obtaining witnesses in his favor, and to have the Assistance of Counsel for his defence.

Amendment VII

In Suits at common law, where the value in controversy shall exceed twenty dollars, the right of trial by jury shall be preserved, and no fact tried by a jury, shall be otherwise reexamined in any Court of the United States, than according to the rules of the common law.

Amendment VIII

Excessive bail shall not be required, nor excessive fines imposed, nor cruel and unusual punishments inflicted.

Amendment IX

The enumeration in the Constitution, of certain rights, shall not be construed to deny or disparage others retained by the people.

Amendment X

The powers not delegated to the United States by the Constitution, nor prohibited by it to the States, are reserved to the States respectively, or to the people.

Note: The capitalization and punctuation in this version is from the enrolled original of the Joint Resolution of Congress proposing the Bill of Rights, which is on permanent display in the Rotunda of the National Archives Building, Washington, D.C.

The Constitution: Amendments 11–27

Constitutional Amendments 1–10 make up what is known as The Bill of Rights.

Amendments 11–27 are listed below.

Amendment XI

Passed by Congress March 4, 1794. Ratified February 7, 1795.

Note: Article III, section 2, of the Constitution was modified by amendment 11.

The Judicial power of the United States shall not be construed to extend to any suit in law or equity, commenced or prosecuted against one of the United States by Citizens of another State, or by Citizens or Subjects of any Foreign State.

Amendment XII

Passed by Congress December 9, 1803. Ratified June 15, 1804.

Note: A portion of Article II, section 1 of the Constitution was superseded by the 12th amendment.

The Electors shall meet in their respective states and vote by ballot for President and Vice-President, one of whom, at least, shall not be an inhabitant of the same state with themselves; they shall name in their ballots the person voted for as President, and in distinct ballots the person voted for as Vice-President, and they shall make distinct lists of all persons voted for as President, and of all persons voted for as Vice-President, and of the number of votes for each, which lists they shall sign and certify, and transmit sealed to the seat of the government of the United States, directed to the President of the Senate; — the President of the Senate shall, in the presence of the Senate and House of Representatives, open all the certificates and the votes shall then be counted; — The person having the greatest number of votes for President, shall be the President, if such number be a majority of the whole number of Electors appointed; and if no person have such majority, then from the persons having the highest numbers not exceeding three on the list of those voted for as President, the House of Representatives shall choose immediately, by ballot, the President. But in choosing the President, the votes shall be taken by states, the representation from each state having one vote; a quorum for this purpose shall consist of a member or members from two-thirds of the states, and a majority of all the states shall be necessary to a choice. [And if the House of Representatives shall not choose a President whenever the right of choice shall devolve upon them, before the fourth day of March next following, then the Vice-President shall act as President, as in case of the death or other constitutional disability of the President. —]* The person having the greatest number of votes as Vice-President, shall be the Vice-President, if such number be a majority of the whole number of Electors appointed, and if no person have a majority, then from the two highest numbers on the list, the Senate shall choose the Vice-President; a quorum for the purpose shall consist of two-thirds of the whole number of Senators, and a majority of the whole number shall be necessary to a choice. But no person constitutionally ineligible to the office of President shall be eligible to that of Vice-President of the United States.

**Superseded by section 3 of the 20th amendment.*

Amendment XIII

Passed by Congress January 31, 1865. Ratified December 6, 1865.

Note: A portion of Article IV, section 2, of the Constitution was superseded by the 13th amendment.

Section 1.

Neither slavery nor involuntary servitude, except as a punishment for crime whereof the party shall have been duly convicted, shall exist within the United States, or any place subject to their jurisdiction.

Section 2.

Congress shall have power to enforce this article by appropriate legislation.

Amendment XIV

Passed by Congress June 13, 1866. Ratified July 9, 1868.

Note: Article I, section 2, of the Constitution was modified by section 2 of the 14th amendment.

Section 1.

All persons born or naturalized in the United States, and subject to the jurisdiction thereof, are citizens of the United States and of the State wherein they reside. No State shall make or enforce any law which shall abridge the privileges or immunities of citizens of the United States; nor shall any State deprive any person of life, liberty, or property, without due process of law; nor deny to any person within its jurisdiction the equal protection of the laws.

Section 2.

Representatives shall be apportioned among the several States according to their respective numbers, counting the whole number of persons in each State, excluding Indians not taxed. But when the right to vote at any election for the choice of electors for President and

Vice-President of the United States, Representatives in Congress, the Executive and Judicial officers of a State, or the members of the Legislature thereof, is denied to any of the male inhabitants of such State, being twenty-one years of age,* and citizens of the United States, or in any way abridged, except for participation in rebellion, or other crime, the basis of representation therein shall be reduced in the proportion which the number of such male citizens shall bear to the whole number of male citizens twenty-one years of age in such State.

Section 3.

No person shall be a Senator or Representative in Congress, or elector of President and Vice-President, or hold any office, civil or military, under the United States, or under any State, who, having previously taken an oath, as a member of Congress, or as an officer of the United States, or as a member of any State legislature, or as an executive or judicial officer of any State, to support the Constitution of the United States, shall have engaged in insurrection or rebellion against the same, or given aid or comfort to the enemies thereof. But Congress may by a vote of two-thirds of each House, remove such disability.

Section 4.

The validity of the public debt of the United States, authorized by law, including debts incurred for payment of pensions and bounties for services in suppressing insurrection or rebellion, shall not be questioned. But neither the United States nor any State shall assume or pay any debt or obligation incurred in aid of insurrection or rebellion against the United States, or any claim for the loss or emancipation of any slave; but all such debts, obligations and claims shall be held illegal and void.

Section 5.

The Congress shall have the power to enforce, by appropriate legislation, the provisions of this article.

**Changed by section 1 of the 26th amendment.*

Amendment XV

Passed by Congress February 26, 1869. Ratified February 3, 1870.

Section 1.

The right of citizens of the United States to vote shall not be denied or abridged by the United States or by any State on account of race, color, or previous condition of servitude—

Section 2.

The Congress shall have the power to enforce this article by appropriate legislation.

Amendment XVI

Passed by Congress July 2, 1909. Ratified February 3, 1913.

Note: Article I, section 9, of the Constitution was modified by amendment 16.

The Congress shall have power to lay and collect taxes on incomes, from whatever source derived, without apportionment among the several States, and without regard to any census or enumeration.

Amendment XVII

Passed by Congress May 13, 1912. Ratified April 8, 1913.

Note: Article I, section 3, of the Constitution was modified by the 17th amendment.

The Senate of the United States shall be composed of two Senators from each State, elected by the people thereof, for six years; and each Senator shall have one vote. The electors in each State shall have the qualifications requisite for electors of the most numerous branch of the State legislatures.

When vacancies happen in the representation of any State in the Senate, the executive authority of such State shall issue writs of election to fill such vacancies: Provided, That the legislature of any State may empower the executive thereof to make temporary appointments until the people fill the vacancies by election as the legislature may direct.

This amendment shall not be so construed as to affect the election or term of any Senator chosen before it becomes valid as part of the Constitution.

Amendment XVIII

Passed by Congress December 18, 1917. Ratified January 16, 1919. Repealed by amendment 21.

Section 1.

After one year from the ratification of this article the manufacture, sale, or transportation of intoxicating liquors within, the importation thereof into, or the exportation thereof from the United States and all territory subject to the jurisdiction thereof for beverage purposes is hereby prohibited.

Section 2.

The Congress and the several States shall have concurrent power to enforce this article by appropriate legislation.

Section 3.

This article shall be inoperative unless it shall have been ratified as an amendment to the Constitution by the legislatures of the several States, as provided in the Constitution, within seven years from the date of the submission hereof to the States by the Congress.

Amendment XIX

Passed by Congress June 4, 1919. Ratified August 18, 1920.

The right of citizens of the United States to vote shall not be denied or abridged by the United States or by any State on account of sex.

Congress shall have power to enforce this article by appropriate legislation.

Amendment XX

Passed by Congress March 2, 1932. Ratified January 23, 1933.

Note: Article I, section 4, of the Constitution was modified by section 2 of this amendment. In addition, a portion of the 12th amendment was superseded by section 3.

Section 1.

The terms of the President and the Vice President shall end at noon on the 20th day of January, and the terms of Senators and Representatives at noon on the 3d day of January, of the years in which such terms would have ended if this article had not been ratified; and the terms of their successors shall then begin.

Section 2.

The Congress shall assemble at least once in every year, and such meeting shall begin at noon on the 3d day of January, unless they shall by law appoint a different day.

Section 3.

If, at the time fixed for the beginning of the term of the President, the President elect shall have died, the Vice President elect shall become President. If a President shall not have been chosen before the time fixed for the beginning of his term, or if the President elect shall have failed to qualify, then the Vice President elect shall act as President until a President shall have qualified; and the Congress may by law provide for the case wherein neither a President elect nor a Vice President shall have qualified, declaring who shall then act as President, or the manner in which one who is to act shall be selected, and such person shall act accordingly until a President or Vice President shall have qualified.

Section 4.

The Congress may by law provide for the case of the death of any of the persons from whom the House of Representatives may choose a President whenever the right of choice shall have devolved upon them, and for the case of the death of any of the persons from whom the Senate may choose a Vice President whenever the right of choice shall have devolved upon them.

Section 5.

Sections 1 and 2 shall take effect on the 15th day of October following the ratification of this article.

Section 6.

This article shall be inoperative unless it shall have been ratified as an amendment to the Constitution by the legislatures of three-fourths of the several States within seven years from the date of its submission.

Amendment XXI

Passed by Congress February 20, 1933. Ratified December 5, 1933.

Section 1.

The eighteenth article of amendment to the Constitution of the United States is hereby repealed.

Section 2.

The transportation or importation into any State, Territory, or Possession of the United States for delivery or use therein of intoxicating liquors, in violation of the laws thereof, is hereby prohibited.

Section 3.

This article shall be inoperative unless it shall have been ratified as an amendment to the Constitution by conventions in the several States, as provided in the Constitution, within seven years from the date of the submission hereof to the States by the Congress.

Amendment XXII

Passed by Congress March 21, 1947. Ratified February 27, 1951.

Section 1.

No person shall be elected to the office of the President more than twice, and no person who has held the office of President, or acted as President, for more than two years of a term to which some other person was elected President shall be elected to the office of President more than once. But this Article shall not apply to any person holding the office of President when this Article was proposed by Congress, and shall not prevent any person who may be holding the office of President, or acting as President, during the term within which this Article becomes operative from holding the office of President or acting as President during the remainder of such term.

Section 2.

This article shall be inoperative unless it shall have been ratified as an amendment to the Constitution by the legislatures of three-fourths of the several States within seven years from the date of its submission to the States by the Congress.

Amendment XXIII

Passed by Congress June 16, 1960. Ratified March 29, 1961.

Section 1.

The District constituting the seat of Government of the United States shall appoint in such manner as Congress may direct:

A number of electors of President and Vice President equal to the whole number of Senators and Representatives in Congress to which the District would be entitled if it were a State, but in no event more than the least populous State; they shall be in addition to those appointed by the States, but they shall be considered, for the purposes of the election of President and Vice President, to be electors appointed by a State; and they shall meet in the District and perform such duties as provided by the twelfth article of amendment.

Section 2.

The Congress shall have power to enforce this article by appropriate legislation.

Amendment XXIV

Passed by Congress August 27, 1962. Ratified January 23, 1964.

Section 1.

The right of citizens of the United States to vote in any primary or other election for President or Vice President, for electors for President or Vice President, or for Senator or Representative in Congress, shall not be denied or abridged by the United States or any State by reason of failure to pay poll tax or other tax.

Section 2.

The Congress shall have power to enforce this article by appropriate legislation.

Amendment XXV

Passed by Congress July 6, 1965. Ratified February 10, 1967.

Note: Article II, section 1, of the Constitution was affected by the 25th amendment.

Section 1.

In case of the removal of the President from office or of his death or resignation, the Vice President shall become President.

Section 2.

Whenever there is a vacancy in the office of the Vice President, the President shall nominate a Vice President who shall take office upon confirmation by a majority vote of both Houses of Congress.

Section 3.

Whenever the President transmits to the President pro tempore of the Senate and the Speaker of the House of Representatives his written declaration that he is unable to discharge the powers and duties of his office, and until he transmits to them a written declaration to the contrary, such powers and duties shall be discharged by the Vice President as Acting President.

Section 4.

Whenever the Vice President and a majority of either the principal officers of the executive departments or of such other body as Congress may by law provide, transmit to the President pro tempore of the Senate and the Speaker of the House of Representatives their written declaration that the President is unable to discharge the powers and duties of his office, the Vice President shall immediately assume the powers and duties of the office as Acting President.

Thereafter, when the President transmits to the President pro tempore of the Senate and the Speaker of the House of Representatives his written declaration that no inability exists, he shall resume the powers and duties of his office unless the Vice President and a majority of either the principal officers of the executive department or of such other body as Congress may by law provide, transmit within four days to the President pro tempore of the Senate and the Speaker of the House of Representatives their written declaration that the President is unable to discharge the powers and duties of his office. Thereupon Congress shall decide the issue, assembling within forty-eight hours for that purpose if not in session. If the Congress, within twenty-one days after receipt of the latter written declaration, or, if Congress is not in session, within twenty-one days after Congress is required to assemble, determines by two-thirds vote of both Houses that the President is unable to discharge the powers and duties of his office, the Vice President shall continue to discharge the same as Acting President; otherwise, the President shall resume the powers and duties of his office.

Amendment XXVI

Passed by Congress March 23, 1971. Ratified July 1, 1971.

Note: Amendment 14, section 2, of the Constitution was modified by section 1 of the 26th amendment.

Section 1.

The right of citizens of the United States, who are eighteen years of age or older, to vote shall not be denied or abridged by the United States or by any State on account of age.

Section 2.

The Congress shall have power to enforce this article by appropriate legislation.

Amendment XXVII

Originally proposed Sept. 25, 1789. Ratified May 7, 1992.

No law, varying the compensation for the services of the Senators and Representatives, shall take effect, until an election of representatives shall have intervened.

Appendix B: Authorizing Statutes for the Armed Forces

U.S. Army

TITLE 10, Subtitle B, PART I, CHAPTER 307.
Sec. 3062. – Policy; composition; organized peace establishment

(a) It is the intent of Congress to provide an Army that is capable, in conjunction with the other armed forces, of –

(1) preserving the peace and security, and providing for the defense, of the United States, the Territories, Commonwealths, and possessions, and any areas occupied by the United States;
(2) supporting the national policies;
(3) implementing the national objectives; and
(4) overcoming any nations responsible for aggressive acts that imperil the peace and security of the United States.

(b) In general, the Army, within the Department of the Army, includes land combat and service forces and such aviation and water transport as may be organic therein. It shall be organized, trained, and equipped primarily for prompt and sustained combat incident to operations on land. It is responsible for the preparation of land forces necessary for the effective prosecution of war except as otherwise assigned and, in accordance with integrated joint mobilization plans, for the expansion of the peacetime components of the Army to meet the needs of war.

(c) The Army consists of –

(1) the Regular Army, the Army National Guard of the United States, the Army National Guard while in the service of the United States and the Army Reserve; and
(2) all persons appointed or enlisted in, or conscripted into, the Army without component.

(d) The organized peace establishment of the Army consists of all –

(1) military organizations of the Army with their installations and supporting and auxiliary elements, including combat, training, administrative, and logistic elements; and
(2) members of the Army, including those not assigned to units; necessary to form the basis for a complete and immediate mobilization for the national defense in the event of a national emergency.

U.S. Marine Corps

TITLE 10, Subtitle C, PART I, CHAPTER 507.
Sec. 5063. – United States Marine Corps: composition; functions

(a) The Marine Corps, within the Department of the Navy, shall be so organized as to include not less than three combat divisions and three air wings, and such other land combat, aviation, and other services as may be organic therein. The Marine Corps shall be organized, trained, and equipped to provide fleet marine forces of combined arms, together with supporting air components, for service with the fleet in the seizure or defense of advanced naval bases and for the conduct of such land operations as may be essential to the prosecution of a naval campaign. In addition, the Marine Corps shall provide detachments and organizations for service on armed vessels of the Navy, shall provide security detachments for the protection of naval property at naval stations and bases, and shall perform such other duties as the President may direct. However, these additional duties may not detract from or interfere with the operations for which the Marine Corps is primarily organized.

(b) The Marine Corps shall develop, in coordination with the Army and the Air Force, those phases of amphibious operations that pertain to the tactics, technique, and equipment used by landing forces.

(c) The Marine Corps is responsible, in accordance with integrated joint mobilization plans, for the expansion of peacetime components of the Marine Corps to meet the needs of war.

U.S. Navy

TITLE 10, Subtitle C, PART I, CHAPTER 507.
Sec. 5062. – United States Navy: composition; functions

(a) The Navy, within the Department of the Navy, includes, in general, naval combat and service forces and such aviation as may be organic therein. The Navy shall be organized, trained, and equipped primarily for prompt and sustained combat incident to operations at sea. It is responsible for the preparation of naval forces necessary for the effective prosecution of war except as otherwise assigned and, in accordance with integrated joint mobilization plans, for the expansion of the peacetime components of the Navy to meet the needs of war.

(b) All naval aviation shall be integrated with the naval service as part thereof within the Department of the Navy. Naval aviation consists of combat and service and training forces, and includes land-based naval aviation, air transport essential for naval operations, all air weapons and air techniques involved in the operations and activities of the Navy, and the entire remainder of the aeronautical organization of the Navy, together with the personnel necessary therefor.

(c) The Navy shall develop aircraft, weapons, tactics, technique, organization, and equipment of naval combat and service elements. Matters of joint concern as to these functions shall be coordinated between the Army, the Air Force, and the Navy.

U.S. Air Force

TITLE 10, Subtitle D, PART I, CHAPTER 807.
Sec. 8062. – Policy; composition; aircraft authorization

(a) It is the intent of Congress to provide an Air Force that is capable, in conjunction with the other armed forces, of –

(1) preserving the peace and security, and providing for the defense, of the United States, the Territories, Commonwealths, and possessions, and any areas occupied by the United States;
(2) supporting the national policies;
(3) implementing the national objectives; and
(4) overcoming any nations responsible for aggressive acts that imperil the peace and security of the United States.

(b) There is a United States Air Force within the Department of the Air Force.

(c) In general, the Air Force includes aviation forces both combat and service not otherwise assigned. It shall be organized, trained, and equipped primarily for prompt and sustained offensive and defensive air operations. It is responsible for the preparation of the air forces necessary for the effective prosecution of war except as otherwise assigned and, in accordance with integrated joint mobilization plans, for the expansion of the peacetime components of the Air Force to meet the needs of war.

(d) The Air Force consists of –

(1) the Regular Air Force, the Air National Guard of the United States, the Air National Guard while in the service of the United States, and the Air Force Reserve;

(2) all persons appointed or enlisted in, or conscripted into, the Air Force without component; and

(3) all Air Force units and other Air Force organizations, with their installations and supporting and auxiliary combat, training, administrative, and logistic elements; and all members of the Air Force, including those not assigned to units; necessary to form the basis for a complete and immediate mobilization for the national defense in the event of a national emergency.

(e) Subject to subsection (f) of this section, chapter 831 of this title, and the strength authorized by law pursuant to section 115 of this title, the authorized strength of the Air Force is 70 Regular Air Force groups and such separate Regular Air Force squadrons, reserve groups, and supporting and auxiliary regular and reserve units as required.

(f) There are authorized for the Air Force 24,000 serviceable aircraft or 225,000 airframe tons of serviceable aircraft, whichever the Secretary of the Air Force considers appropriate to carry out this section. This subsection does not apply to guided missiles.

U.S. Coast Guard

TITLE 14, PART I, CHAPTER 1.
Sec. 1. – Establishment of Coast Guard

The Coast Guard as established January 28, 1915, shall be a military service and a branch of the armed forces of the United States at all times. The Coast Guard shall be a service in the Department of Homeland Security, except when operating as a service in the Navy.

Sec. 2. – Primary duties

The Coast Guard shall enforce or assist in the enforcement of all applicable Federal laws on, under, and over the high seas and waters subject to the jurisdiction of the United States; shall engage in maritime air surveillance or interdiction to enforce or assist in the enforcement of the laws of the United States; shall administer laws and promulgate and enforce regulations for the promotion of safety of life and property on and under the high seas and waters subject to the jurisdiction of the United States covering all matters not specifically delegated by law to some other executive department; shall develop, establish, maintain, and operate, with due regard to the requirements of national defense, aids to maritime navigation, ice-breaking facilities, and rescue facilities for the promotion of safety on, under, and over the high seas and waters subject to the jurisdiction of the United States; shall, pursuant to international agreements, develop, establish, maintain, and operate icebreaking facilities on, under, and over waters other than the high seas and waters subject to the jurisdiction of the United States; shall engage in oceanographic research of the high seas and in waters subject to the jurisdiction of the United States; and shall maintain a state of readiness to function as a specialized service in the Navy in time of war, including the fulfillment of Maritime Defense Zone command responsibilities.

Appendix C: Service Values of the Armed Forces

U.S. Army
Loyalty
Duty
Respect
Selfless Service
Honor
Integrity
Personal Courage

U.S. Navy and Marine Corps
Honor
Courage
Commitment

U.S. Air Force
Integrity First
Service Before Self
Excellence in All We Do

U.S. Coast Guard
Honor
Respect
Devotion to Duty

Appendix D: Code of Conduct for Members of the United States Armed Forces

I

I am an American, fighting in the forces which guard my country and our way of life. I am prepared to give my life in their defense.

II

I will never surrender of my own free will. If in command, I will never surrender the members of my command while they still have the means to resist.

III

If I am captured I will continue to resist by all means available. I will make every effort to escape and aid others to escape. I will accept neither parole nor special favors from the enemy.

IV

If I become a prisoner of war, I will keep faith with my fellow prisoners. I will give no information or take part in any action which might be harmful to my comrades. If I am senior, I will take command. If not, I will obey the lawful orders of those appointed over me and will back them up in every way.

V

When questioned, should I become a prisoner of war, I am required to give name, rank, service number and date of birth. I will evade answering further questions to the utmost of my ability. I will make no oral or written statements disloyal to my country and its allies or harmful to their cause.

VI

I will never forget that I am an American, fighting for freedom, responsible for my actions, and dedicated to the principles which made my country free. I will trust in my God and in the United States of America.

Executive Order 10631 (1955) as amended by EO 11382 (1967) and EO 12633 (1988)

About the Authors

DR. ALBERT C. PIERCE is the National Defense University's professor of Ethics and National Security. He also worked on the 2007 edition of this work.

DR. RICHARD M. SWAIN is a retired Army colonel and former professor of Officership at the U.S. Military Academy at West Point. He also worked on the 2007 edition of this work.